Narrative Form and Chaos Theory in Sterne, Proust, Woolf, and Faulkner

Also by Jo Alyson Parker

The Author's Inheritance: Henry Fielding, Jane Austen, and the Establishment of the Novel

Time and Memory: The Study of Time XII (coedited with Michael Crawford and Paul Harris)

Narrative Form and Chaos Theory in Sterne, Proust, Woolf, and Faulkner

Jo Alyson Parker

NARRATIVE FORM AND CHAOS THEORY IN STERNE, PROUST, WOOLF, AND FAULKNER
Copyright © Jo Alyson Parker, 2007.

Reprinted Jo Alyson Parker: "Strange Attractors in *Absalom, Absalom!*" from *Reading Matters: Narrative in the New Media Ecology*, edited by Joseph Tabbi and Michael Wutz. Copyright © 1997 by Cornell University. Used by permission of the publisher, Cornell University Press.

Reprinted Jo Alyson Parker: "'The Clockmakers Outcry': *Tristram Shandy* and the Complexification of Time" from *Disrupted Patterns: On Chaos and Order in the Enlightenment*, edited by Theodore E. D. Braun and John McCarthy. Copyright © 2000 Rodopi. Used by permission of the publisher, Rodopi.

Figures generated by Thomas Weissert used by his permission.

All rights reserved. No part of this book may be used or reproduced in any manner whatsoever without written permission except in the case of brief quotations embodied in critical articles or reviews.

First published in 2007 by
PALGRAVE MACMILLAN™
175 Fifth Avenue, New York, N.Y. 10010 and
Houndmills, Basingstoke, Hampshire, England RG21 6XS
Companies and representatives throughout the world.

PALGRAVE MACMILLAN is the global academic imprint of the Palgrave Macmillan division of St. Martin's Press, LLC and of Palgrave Macmillan Ltd. Macmillan® is a registered trademark in the United States, United Kingdom and other countries. Palgrave is a registered trademark in the European Union and other countries.

ISBN-13: 978–1–4039–8384–8
ISBN-10: 1–4039–8384–4

Library of Congress Cataloging-in-Publication Data is available from the Library of Congress.

A catalogue record for this book is available from the British Library.

Design by Newgen Imaging Systems (P) Ltd., Chennai, India.

First edition: September 2007

10 9 8 7 6 5 4 3 2 1

Printed in the United States of America.

PN
212
.P37
2007

To Tom

Contents

List of Figures	ix
Preface	xi
Acknowledgments	xv
1 Chaos Theory and the Dynamics of Narrative	1
2 Narrating against the Clockwork Hegemony: *Tristram Shandy*'s Games with Temporality	31
3 Narrating the Workings of Memory: Iteration and Attraction in *In Search of Lost Time*	61
4 Narrating the Unbounded: Mrs. Dalloway's Life, Septimus's Death, and Sally's Kiss	87
5 Narrating the Indeterminate: Shreve McCannon in *Absalom, Absalom!*	111
Postscript	131
Notes	135
Bibliography	169
Index	183

List of Figures

1.1	Fixed-point attractor	13
1.2	Periodic orbit	14
1.3	A Rössler or funnel strange attractor	15
1.4	A Lorenz or butterfly strange attractor	16
3.1	A butterfly strange attractor	84
3.2	A butterfly strange attractor collapsing into a fixed-point attractor	84

Preface

In Tom Stoppard's witty and elegant play *Arcadia*, the nineteenth-century math prodigy Thomasina discovers the disorderly order of deterministic chaos that practitioners of chaos theory investigate today. She lacks, however, the mathematical language to articulate her findings, and, perhaps more important, she lacks the tool—the superfast computer—that would do the innumerable mathematical iterations necessary to display them. Thomasina burns to death on the night before her seventeenth birthday, and Septimus, her tutor and would-be lover, ends up as the lunatic Hermit of Sidley Park, performing those endless iterations in a tribute to his lost love. Thus, in Stoppard's excursion into the realm of might-have-been, Thomasina's possibly paradigm-shifting discovery is lost, relegated to a few disregarded notes in a schoolgirl's copybook and the seemingly meaningless calculations of a madman. As Valentine, the modern-day biologist who validates Thomasina's discovery, explains, "You can't open a door till there's a house."[1] Stoppard's play thus dynamically demonstrates that, for our ideas to get a hearing, a felicitous convergence of events must occur.

My project owes its genesis to such a felicitous convergence. In the nearly two hundred years since the fictional Thomasina and Septimus danced their first and final waltz, the "house" was built, and the "door" could be opened onto the vista of deterministic chaos. Because of computer technology, dynamicists can now perform the iterations that enable them to discern such chaos. Granted, such iterations can be done without the aid of the computer, but the process would be so time-consuming as to make it unfeasible—the work of a mad hermit.[2] Computer technology enables simulation and, consequently, allows dynamicists to apprehend and articulate the indeterminate determinism common to certain chaotic dynamical systems.

At the same time that the "new science" of chaos was generating a buzz, an interest in establishing connections between literature and science pervaded the academy.[3] The Society for Literature and Science (SLS—now the Society for Literature, Science, and the Arts) was founded in 1985, and it has grown rapidly since then. Its annual meeting and journal *Configurations* testify to the significance, applicability, and popularity of making connections between the humanities and the sciences, including applications of chaos theory to literary studies. The SLS annual meeting, in fact, provided the forum for my early work in this area, and the panel presentations and subsequent discussions supplied a fertile ground for helping me develop my ideas.

The interest in literary applications of chaos theory has gone beyond what might appear to be the specialized focus of SLS members. Annual meetings of such organizations as the Modern Language Association, the Society for the Study of Narrative Literature, and the American Society for Eighteenth-Century Studies have offered sessions that deal with the implications of chaos theory for literary studies. The 1995 meeting of the interdisciplinary International Society for the Study of Time was devoted to the subject of deterministic chaos, including its applications for literature. The Society for Chaos Theory in Psychology and Life Sciences features literary topics at its annual conference, further testimony to the interdisciplinary attraction of chaos theory. Major literary journals, including *New Literary Theory*, *PMLA*, and *Poetics Today*, feature essays on the subject, and significant full-length studies have explored literary applications of chaos theory.[4] In the approximately twenty years since chaos theory seized the public imagination, it has demonstrated real staying power, not only in the sciences but also in cultural studies.

In the 1990s, when I began teaching a seminar course in narrative form, I found myself applying the insights yielded by chaos theory to my readings of certain paradigmatic narratives. Each time that I taught the course, I benefited from the insights of a group of committed and lively students. Together, we focused on a variety of texts that foregrounded their own narrative dynamics. Chaos theory regularly provided a means of understanding these dynamics.

Out of this Zeitgeist, the following text has emerged. Drawing on contemporary theories of dynamical systems and of narrative, it melds theory and practice. It thus features two parts: (1) a detailed theoretical introduction and (2) readings of particular texts whose structure mimics that of chaotic systems. Overall, parts one and two function together as a feedback loop in that chaos theory sheds new light on the narratives and the texts, in turn, make concrete the abstractions of the theory.

The four texts that I explore are all are novels from the modern age, so I use the term narrative deliberately. To focus on the novel as such invites a discussion of its generic features and the general social, cultural, and historical circumstances out of which it arose during the eighteenth century. I choose to emphasize a dynamical structure rather than a genre and to demonstrate how each of the four texts presents a particular chaotic response to a particular narrative problem. Because the novel genre is the most important form of sustained narrative, it offers rich veins for exploration.

I should make clear that I do not put forward a history or progress of chaotic narrative. I do, however, arrange the texts chronologically, moving from the mid-eighteenth to the mid-twentieth century, and I suggest how earlier writers may have influenced later ones. As I argue, the four texts serve as exemplary chaotic narratives, and examining them through the lens of chaos theory illuminates their complex dynamics.

Acknowledgments

Many people have helped me bring this book to fruition. My initial interest in narrative form was sparked by a long-ago course taught by Alexander Gelley, whose insights helped me begin working toward my own theory of narrative. Owen Gilman encouraged me to return to earlier work on *Absalom, Absalom!*, and the result was some of my first work in chaos theory. Michael Wutz and Joseph Tabbi provided me with my first opportunity to publish such work. Theodore E. D. Braun's panel on "eighteenth-century chaos" at the meeting of the American Society for Eighteenth-Century Studies prompted me to tackle *Tristram Shandy*, and he and John McCarthy spurred me to examine the novel further for the collection *Disrupted Patterns*. Alexander Argyros, Maria Assad, J. T. Fraser, and Paul Harris helped enlarge my thinking on narrative form and chaos theory.

Early versions of the book chapters profited from the comments of Saint Joseph's University English faculty who participated in the summer writing group, including Thomas Brennan, S. J., Melissa Goldthwaite, Ann Green, Richard Haslam, Nimisha Ladva, April Lindner, and Deborah Scott. Richard Fusco read the entire manuscript through at least twice, and his editorial comments have proven invaluable. I am grateful to Eileen Cohen, Elizabeth Doherty, Jane Fraser, Virginia Johnson, Francis Morris, Audra Parker, Kevin Parker, and Katherine Sibley for the support and encouragement that they provided during the writing process. Loretta Giello's administrative abilities consistently have helped make onerous tasks more bearable. I would also like to thank the students of my "Seminar in Narrative Form" for their provocative questions and perceptive comments.

The Saint Joseph's University Board on Faculty Research and Development provided me with a summer grant and a sabbatical leave, both of which helped me to complete this book. I wish to thank Farideh Koohi-Kamali and Julia Cohen of Palgrave Macmillan for their advice and support during the publication process.

Finally, my greatest thanks go to my family. At the age of four, my daughter Lizzy defined the exemplary narrative plot: "Once upon a time. The middle. The end." Now in her teens, she fulfills the potential of this early wisdom in her increasingly more sophisticated analyses of narrative, which help me refine my own thinking. She has been patient and encouraging as I have worked through the various drafts of the manuscript. I am grateful to my husband, Thomas Weissert, for support both personal and professional. Our conversations about narrative form and chaos theory began many years ago, and they inspired me to begin writing. Tom's rigorous reading of the manuscript has helped me, a nonscientist, refine my thinking. Any scientific gaffes rest with me alone. I thank him also for generating the figures that appear in chapters 1 and 3.

Parts of chapter 2 were published as "Spiraling down 'the Gutter of Time': *Tristram Shandy* and the Strange Attractor of Death" in *Weber Studies* 14 (1997): 102–14 and as "'The Clockmakers Outcry': *Tristram Shandy* and the Complexification of Time" in *Disrupted Patterns: On Chaos and Order in the Enlightenment*, edited by Theodore E. D. Braun and John McCarthy (Amsterdam-Atlanta, GA: Rodopi, 2000.): 147–60. An early version of chapter 5 was published as "Strange Attractors in *Absalom, Absalom!*" in *Reading Matters: Narrative in the New Media Ecology*, edited by Joseph Tabbi and Michael Wutz (New York: Cornell University Press, 1997) 99–118. I thank the publishers for permission to draw upon this material.

CHAPTER 1

Chaos Theory and the Dynamics of Narrative

This is not science. This is story-telling.
—Tom Stoppard, *Arcadia*

Science looked a lot like literary criticism, from across the room.
—Richard Powers, *Galatea 2.2*

But attractors are themselves models. They are metaphors for processes.
—J. T. Fraser, "From Chaos to Conflict"

What you are about to read is not science but storytelling, a narrative about narrative. In the following pages, I examine how science may indeed look "a lot like literary criticism." Specifically, I look at how contemporary ways of modeling turbulent dynamical systems in the physical world look like models of a certain kind of literary narrative structure, and I consider what the implications of that analogy are.

During the early 1980s, with the aid of computer-generated simulations, scientists discovered or, more accurately, identified deterministic chaos, a circumstance that "has created a new paradigm in scientific modeling," according to four of the founders of chaos theory.[1] Chaos theory enables us to see the physical world in new ways and to look anew at texts that I call "chaotic." By viewing such texts through a chaos-theory lens, we can link narrative structure with narrative content and link the formalism of traditional narratology with the reader's production of narrative meaning.

"Chaos theory" is a nonspecialist, catchall term that we use to cover the study of chaotic behaviors in a variety of disciplines—biology, chemistry, economics, and so forth. I focus on chaos theory as practiced within physics, wherein it occupies a particular subdiscipline called "dynamical systems theory."[2] Although "dynamical systems theory" more accurately designates the scientific modeling I hereafter describe, I nevertheless use "chaos theory" to acknowledge the phrase's greater cultural resonance.[3]

Chaos theory has changed the way in which we conceptualize so-called chaotic structures in the natural world. Once regarded as "poor in order," chaos has come to be seen as "rich in information," according to N. Katherine Hayles, one of the first literary scholars to draw on chaos theory.[4] Once seen as aberrant, the nonlinear and the random are now understood as prevalent, and physical behaviors once disregarded and dismissed are now considered legitimate areas of inquiry. The most far-reaching insights that chaos theory offers us are that patterns of order emerge spontaneously out of random behavior, that deterministic systems can generate random behavior when small uncertainties are amplified as the system develops through time, and that time itself can operate differently at local levels. Models of chaotic systems demonstrate the entanglement of system and systematizer in generating meaning, a feedback loop thus running between the subjective observer and the object under observation.

By looking through a chaos-theory lens, we can gain new insights into narratives whose structures display chaotic qualities. Such a reading enables us to apprehend how their form is their meaning, which emerges from the particular social, cultural, and historical circumstances, and how their meaning is dynamical, entangling the reader in the interpretive process. Through the perspective afforded us by chaos theory, we can discern the disorderly order—the complex yet simple elegance—of these narratives.

The Ordered Universe of Classical Physics

In *Tristram Shandy*, Tristram's Uncle Toby, engaging in a hobby-horsical attempt to recreate the Battle of Namur, calculates the trajectories of the cannonballs that were fired. If he knows the position and velocity of the cannonballs at a certain time, he should be able to predict their future state. The episode thus demonstrates the way in which the deterministic, time-symmetric assumptions of Newtonian or classical physics enable one to solve a particular sort of physical problem and thus attain an apparent mastery of the physical world.

From the time Uncle Toby appeared in fiction until the late twentieth century, what Julian Hunt called the "Newtonian-Laplacian clockwork view of the universe" held sway in the sciences.[5] Stephen Kellert aptly characterizes these beliefs as "the clockwork hegemony."[6] According to the Newtonian paradigm, we live in a universe whose workings function as regularly and predictably as those of a clock (an infallible, perpetual clock), its hands sweeping across its face in an exactly repeatable motion and at exactly the same rate of speed. Classical physics is predicated on the related notions of stability, repeatability, predictability, causality, absolute time, and observer objectivity.

Classical physics focuses on a class of physical systems whose entire behavior can be exactly calculated with a set of equations. Peter Covenay and Roger Highfield explain their predictive power: "Newton's equations of motion are such that, no matter what the positions and velocities at an initial time of observation—the *initial conditions*—the behavior of the system is determined for all future and past times" (emphasis in the original).[7] For instance, a frictionless pendulum always obtains an exactly repeating cycle, even when we start it swinging in a different way each time. Its behavior is predictable. James Crutchfield et al. claim, "The great power of science lies in the ability to relate cause and effect. On the basis of the laws of gravitation, for example, eclipses can be predicted thousands of years in advance."[8] Clearly, classical physics explains and predicts with accuracy many physical systems.

However, classical physics also ignores those systems that cannot be accurately predicted. Stephen Kellert comments upon this "prejudice": "Education in the natural sciences created the impression that linear and solvable systems were the only ones (or at least the only important ones)."[9] The case of Edward Lorenz is exemplary. In 1963, he published what later came to be regarded as groundbreaking chaos-theory articles in meteorology journals, which physicists ignored.[10] The classical physicist examines the dynamics of a pendulum or the solar system but not those of the weather or a dripping faucet or a water wheel.

The essence of classical physics resides in the reflexivity of predictability and determinism: if a system's behavior is predictable, the system is deterministic, and if a system is deterministic, its behavior is predictable. Because of this exclusive focus on predictable systems, classical physics leads to an inherently deterministic view of the universe. Indeed, in the early nineteenth century, Pierre Simon de Laplace envisioned an imaginary entity—a "demon"—who would be "capable at any given instant of observing the position and velocity of each mass that forms part of the universe and inferring its evolution, both toward the past and toward the future."[11] This demon could retrodict all past states of the universe and predict all future ones; it was "an intelligence that recognizes all forces of nature and the elements that compose it," for whom "nothing would be uncertain."[12] Until the advent of chaos theory, science worked on the assumption that such a "demonic" intelligence could be achieved, enabling, as Julian Hunt suggests, the ultimate control of nature: "One might describe the mid-twentieth century view as the confident belief that the natural world is largely predictable and rational, so that with the assistance of information theory, computing power, and system control . . . it would be possible even for the natural world to be controlled by human intervention."[13] According to the classical paradigm, with the right tools, we could eventually fathom all the workings of the universe—a situation that, as Ivar Ekeland wryly observes, "is enough to stifle with boredom several generations of astronomers."[14]

If we assume that the workings of the universe are completely predictable, running like a well-regulated clock, certain assumptions about time pertain. In order to predict future events and retrodict past ones accurately, we must assume that a fixed rate of time pertains. The Newtonian view of time is, after all, absolutist, predicated on notions of linearity and periodicity. In *Principia Mathematica*, Newton makes his well-known distinction between absolute time and relative time:

> Absolute, true, and mathematical time, of itself, and from its own nature, flows equably without relation to anything external, and by another name is called duration: relative, apparent, and common time, is some sensible and external (whether accurate or unequable) measure of duration by the means of motion, which is commonly used instead of true time; such as an hour, a day, a month, a year.[15]

According to this absolutist view of time, just as we see events as occurring in a definite place, we see them also as occurring at a definite time, as Ilya Prigogine and Isabelle Stengers explain, "In classical mechanics time was a number characterizing the position of a point on its trajectory."[16] In essence, we assume that events can be fixed upon a uniform time line.

Interestingly, in the passage from the *Principia*, Newton discriminates between an idealized time independent of any external factors and a timing of time, which involves using an external means of measuring it. Michel Serres observes, however, "*People usually confuse time and the measurement of time,* which is a metrical reading on a straight line" (emphasis in the original).[17] Such confusion stems from an implicit connection between the mechanical clock and the Newtonian notion of absolute time as a laminar flow, unchanging in its rate. According to G. J. Whitrow, the mechanical clock may actually have given rise to the notion of uniform time:

> [T]he invention of an accurate mechanical clock had a tremendous influence on the concept of time itself. For unlike the clocks that preceded it, which tended to be irregular in their operation, the improved mechanical clock when properly regulated could tick away uniformly and continually for years on end, and so must have greatly strengthened belief in the homogeneity and continuity of time. The mechanical clock was therefore not only the prototype instrument for the mechanical conception of the universe but for the modern idea of time.[18]

The clock becomes, in Prigogine and Stengers's terms, "the very symbol of world order."[19] God is regarded as the divine watchmaker who wound up the great machine of the universe and left it to tick away at a regular, predictable rate.

The clockwork view of the universe depends on the notion of observer objectivity—that is, the notion that the observer merely records but does not shape natural phenomena. Evelyn Fox Keller points out the connection that is made between observer objectivity and scientific progress: "It is often argued that the very success of modern science and technology rests on a new methodology that protects its inquiries from the idiosyncratic sway of human motivation."[20] An integral premise of the scientific method is that scientists, irrespective of their particular situation, will obtain the same results when performing the same experiment—the agent who performs the action thus nonessential to the results. This premise of scientific objectivity sets up a separation between nature and humanity, as Priogogine and Stengers describe: "Man is emphatically not part of the nature he objectively describes; he dominates it from the outside."[21]

The clock ticks away, and all that the observer does, in fact, is to observe and record.

Divested of its moral aims, the following passage from Alexander Pope's *An Essay on Man* sums up the ordered, deterministic Newtonian cosmos:

> All nature is but Art, unknown to thee;
> All Chance, Direction, which thou canst not see;
> All Discord, Harmony, not understood;
> All partial Evil, universal Good. (1.289–92)[22]

Pope, of course, asserts that our limited understanding precludes our ability to discern this grand design in full. Classical physics, however, with its focus on the periodic and stable, presumes that we might eventually be able to do so.

Challenging the "Clockwork Hegemony"

Although chaos theory undermines our conception of an inherently ordered universe whose workings might eventually be apprehended in full, it does not substitute a contrary notion that an absence of any order pertains to the universe—nor could it, considering the explanative efficacy of the Newtonian paradigm. Instead, chaos theory spotlights the paradoxically termed "deterministic chaos" and "order out of chaos" that exist in the physical world. As Ian Stewart succinctly sums up, such chaos is "lawless behavior governed entirely by law."[23] How might this paradox hold true? It does so because chaos provides a way of modeling systems whose behavior, although subject to deterministic physical laws, nonetheless exhibit unpredictable behavior. Alexandre Favre explains the paradox thus: "The fact that a complex phenomenon cannot be predicted exactly in the long term with the techniques now available is not incompatible with its being fully determined by the principles of physics."[24] Significantly, with regard to certain kinds of dynamical systems, chaos theory severs the connection that classical physics makes between determinism and predictability.

Until the late twentieth century, the clockwork hegemony initiated by classical physics led scientists to dismiss dynamical systems whose behavior was not predictable, such as weather or—a more modest object of study—a dripping faucet with an increased flow rate. Kellert claims that "the appeal of stable periodic motion was somehow so great that physicists began to see everything as a clock, to the extent that nonperiodic behavior was denied or dismissed."[25] Certainly, scientists were aware that such unpredictable behavior occurred in the physical world, but they saw it as aberrant. Over the past several decades, however, systems whose behavior cannot be predicted have become significant objects of study as scientists acknowledge the prevalence of deterministic chaos in the natural world.

In the early 1960s, Edward Lorenz used nonlinear differential equations to represent the atmosphere as a simple convection system. In the resultant paper, provocatively titled "Deterministic Nonperiodic Flow," Lorenz focused on "nonperiodic solutions, i.e., solutions which never repeat their past history exactly, and where all approximate repetitions are of finite duration."[26] Differential equations are deterministic, but, as Lorenz discovered, the solutions were not necessarily determinable, a conclusion with far-reaching ramifications for long-range weather forecasting:

> When our results concerning the instability of nonperiodic flow are applied to the atmosphere, which is ostensibly nonperiodic, they indicate that prediction of the sufficiently distant future is impossible by any method, unless the present conditions are known exactly. In view of the inevitable inaccuracy and incompleteness of weather observations, precise very-long-range forecasting would seem to be non-existent.[27]

We can determine that the summer temperature in, say, Philadelphia will range in the 80s and 90s, but we cannot predict when a thunderstorm will occur or how fierce it will be.

Although Lorenz's "paper languished in obscurity" for a good ten years, it heralded the swerve toward examining deterministic chaos.[28]

In 1984, Robert Shaw published *The Dripping Faucet as a Model Chaotic System*, which serves as an exemplary study of deterministic chaos on a small scale. The faucet exemplifies "a system capable of a *chaotic transition*": that is, a system that "can change from a periodic and predictable to an aperiodic quasi-random pattern of behavior, as a single parameter (in this case, the flow rate) is varied."[29] At a certain flow rate, the drips may be regular, predictable. But the system is nonlinear: when a drop detaches from the string of water, a sudden change in mass occurs, and each drop affects a subsequent drop. If we turn up the flow rate, the system goes chaotic as the shortened time between drops makes their interaction unpredictable.[30] James Gleick discusses the significance of this behavior in his chapter on the Dynamical Systems Collective:

> The interesting feature of the model—the *only* interesting feature, and the nonlinear twist that made chaotic behavior possible—was that the next drip depended on how the springiness interacted with the steady increasing weight. A down bounce might help the weight reach the cutoff point that much sooner, or an up bounce might delay the process slightly.[31]

Certainly, the behavior of the drops is deterministic; laws of classical physics guarantee that they *will* fall and that their mass and rate of speed in falling *will* lie within certain boundaries. We cannot, however, solve the equations that would enable accurate predictions of when, where, and how the drops will fall.[32] As both Lorenz and Shaw demonstrated, with a chaotic dynamical system there are too many variables for which we cannot account, no matter how precise our instruments are.[33] Hence deterministic chaos.

Our inability to predict precisely the evolution of a chaotic system is due to its sensitive dependence on initial conditions, whereby "microscopic perturbations are amplified to affect macroscopic behavior"—the so-called butterfly effect.[34] Crutchfield et al. illustrate this effect with an example of a billiard player "with perfect control over his or her stroke" trying to predict the trajectory of the ball: "If the player ignored an effect even as miniscule as the gravitational attraction of an electron at the edge of the galaxy, the prediction would become wrong after one minute!"[35] The slightest variation in variables can lead to exponentially different result. When Lorenz, for example, rounded off the digits specifying initial conditions in an attempt to replicate earlier results, he obtained very different results over a period of time: "[T]wo states differing by imperceptible amounts may eventually evolve into two considerably different states."[36]

Because we cannot predict the causal connection between a past state and a future one in chaotic systems, we must rethink our assumptions about time. Newton's notion of an absolute time, apparently running along a symmetrical line, cannot help us account for the temporal qualities of a chaotic system.[37] J. T. Fraser, one of the foremost philosophers of time, discusses the emergence of "new kinds of temporal qualities" in light of chaos theory: "[T]ime ceases to be a background to events and is understood, instead, as constituting an evolving aspect of reality and a correlate of complexity. . . . the evolution of time is not a progress into preexisting forms of time but the creative emergence of increasingly more complex temporalities."[38] Rather than regarding time as operating symmetrically according to some invariant global law, we can begin to regard it as operating differently, yet no less validly, at different local levels.[39]

In addition to making us reconsider our notions of time, chaos theory also prompts us to rethink our notions of observer objectivity. I do not mean that science thus becomes idiosyncratically subjective. Chaotic phenomena are subject to objective mathematical description, just as the phenomena examined by classical physics are. Nevertheless, the observer becomes acknowledged as an integral part of the meaning-making process.⁴⁰ In a 1981 Stanford symposium on order and disorder, Edgar Morin describes this interconnection:

> Randomness opens up the uncertain problematic of the human mind confronted at once with reality and its own reality. The old determinism was an ontological affirmation [of] the nature of reality. Randomness introduces a relationship between observer and reality. The old determinism excluded the organization, the environment, the observer.⁴¹

According to Morin, "the real field of knowledge is not the pure object, but the object viewed, perceived, and co-produced by us. . . . In other words, the object of knowledge is phenomenology and not ontological reality."⁴² Through his description of a dynamicist engaged in mapping chaotic behavior on a computer screen, Thomas Weissert enables us to understand how the observer coproduces the object under observation:

> Every so often, he would type a key and so bump the control parameter: the scene changed, eddies appeared, and the swirling became more intense. . . . This reality was the abstract Cartesian space where points and lines represented evolving dynamical systems. Yet by immersing his sense, sight and touch, the dynamicist was climbing into phase space and taking it for a ride.⁴³

Through actively manipulating the system parameters, the dynamicist produces the simulation that represents the system. The observer no longer merely observes but provides a context for our understanding of the object under observation.

Strange Attractors and Fractals

Two closely related manifestations of deterministic chaos have particular significance for my argument about narrative: the strange attractor and the fractal. We can think of the strange attractor as a behavior of a chaotic system and a fractal as a geometrical property of the system with the latter characterizing the geometry of the former.

When we simulate the behavior of dynamical systems on a computer, we can discern a clear distinction between those that exhibit classically deterministic behavior and those that exhibit chaotic behavior.[44] We map such behavior along the Cartesian coordinates of what is called "state space," wherein each point represents one possible unique configuration of the system at a particular point in time. The figure thereby generated represents the dynamical system's evolution numerically. An attractor is simply what its name suggests: "what the behavior of a system settles down to, or is attracted to," as shown by its evolution in state space.[45] The identity of each attractor comprises, as Weissert explains, "the union of all initial conditions whose trajectories arrive there; that union is called the *basin of attraction* for that attractor" (emphasis in the original).[46] To understand this notion, we might literalize the metaphor and think back to Shaw's serendipitous dripping faucet; although we cannot know where and when all those annoying drops will fall, we do know that they will stay in the basin.

If the trajectory of the dynamical system assumes a repeating pattern in state space, we say that it has reached the attractor. For example, reconsider the pendulum. When mapping its behavior in state space, we observe a trajectory that spirals inward to a fixed point, the figure thus representing the pendulum's attraction to a final state of stasis (see figure 1.1.). When mapping the behavior of a frictionless pendulum, we observe a periodic orbit, the figure thus representing the pendulum's attraction to a particular set of repeating coordinates (see figure 1.2.). Systems such as pendulums exhibit classically deterministic behavior because, no matter how we vary the initial conditions within the basin of attraction, the trajectory will always fall onto the same attractor.

When we represent the evolution of a chaotic dynamical system, such as a water wheel or the weather, we often observe a strange attractor. Consider again the dripping faucet. A computer simulation of its behavior gives us access to the three variables of the drops' position, velocity, and mass over

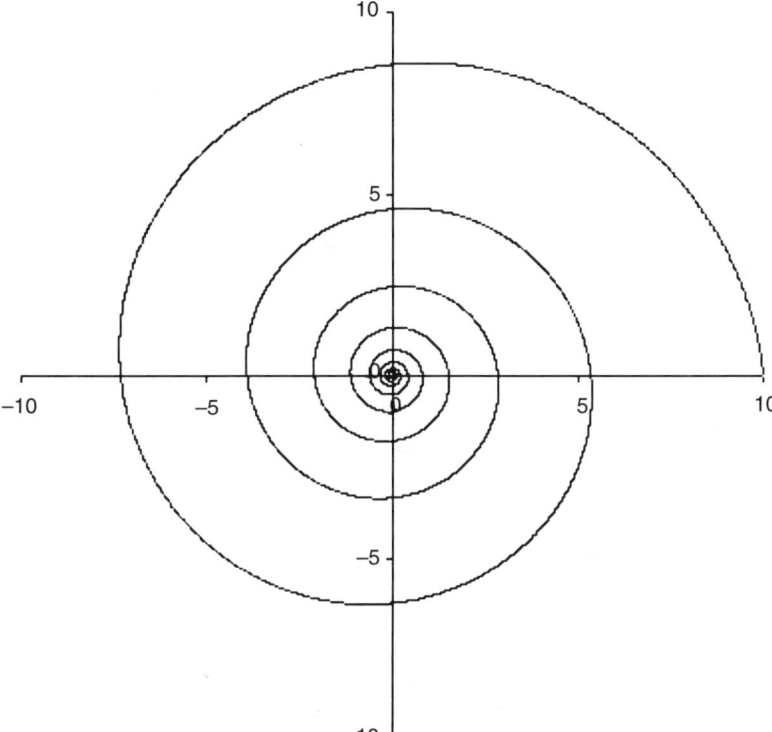

Figure 1.1 Fixed-point attractor

an indefinite period of time and thus enables us to see, in Shaw's words, "the geometry of the 'attractor' describing the motion of the fluid system."[47] We find that the orbit hovers around certain coordinates within a basin of attraction, but we cannot predict exactly when the orbit will move closer or farther away from those coordinates. Trajectories diverge at times and almost converge at other times, but they never repeat themselves exactly, the evolving shape constituting a Rössler or funnel strange attractor (see figure 1.3.). Crutchfield et al. point out that "in a typical chaotic transformation recurrence is exceedingly rare, occurring perhaps only once in the lifetime of the universe."[48] State space might fill up completely with orbits,

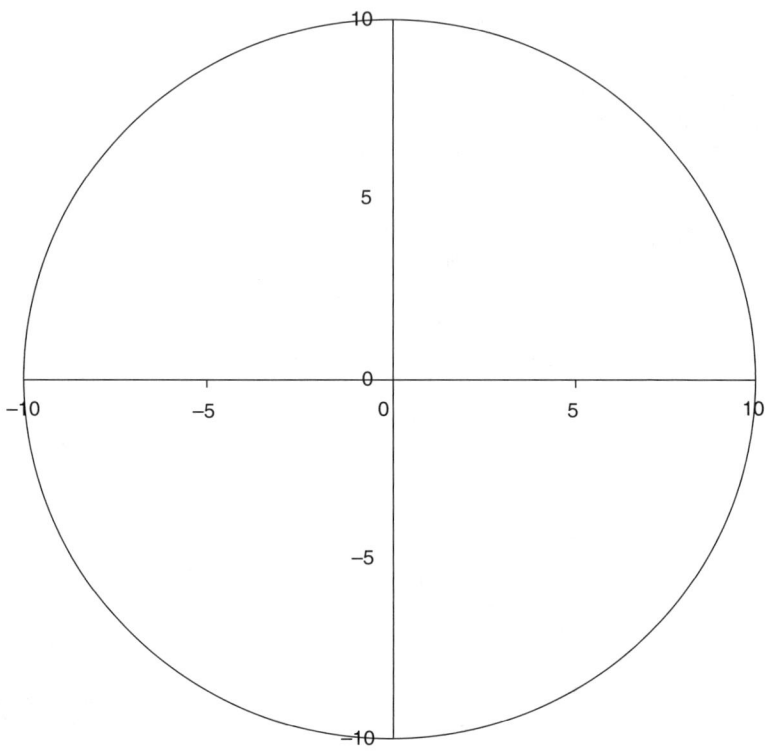

Figure 1.2 Periodic orbit

and the attractor (the ultimate fate of the system) might reveal itself—had we but world enough and time.

In a chaotic system, there may be an actual attracting point (or points, depending on the system). Unlike the stable fixed point upon which the pendulum eventually comes to rest, however, the attracting point in a chaotic system has become unstable. Michael Berry describes the state-space portrait of what he calls "the bouncer" toy (a contraption featuring magnetized swinging balls), which has an unstable attracting point: "[O]ur rotator never comes to rest . . . in spite of being continually damped by friction, because it is being *driven* by the magnet through the swinging of the heavy pendulum. Therefore its stroboscopic phase portrait must be generated by an area-shrinking

Figure 1.3 A Rössler or funnel strange attractor

map whose attractor is more complicated than a single point representing rest."[49] The attracting point concurrently attracts and repels the system trajectory, ensuring that the trajectory will never actually pass through it but come closer only to veer away.

In the case of a Lorenz or butterfly attractor (a particular subset of strange attractors), the orbit jumps between two attracting points, and we cannot predict when the jump will occur; the trajectory "crosses from one spiral to the other at irregular intervals"[50] (see figure 1.4.) Comprising an infinite amount of local variations within fixed global limits, strange attractors are determinate in their global spatial-temporal patterning and indeterminate in their precise evolution. Through their qualities of bounded randomness, they concurrently model deterministic chaos and serve as apt figures of it.[51]

Although a page of paper such as this can show only a two-dimensional and static representation of a strange attractor,

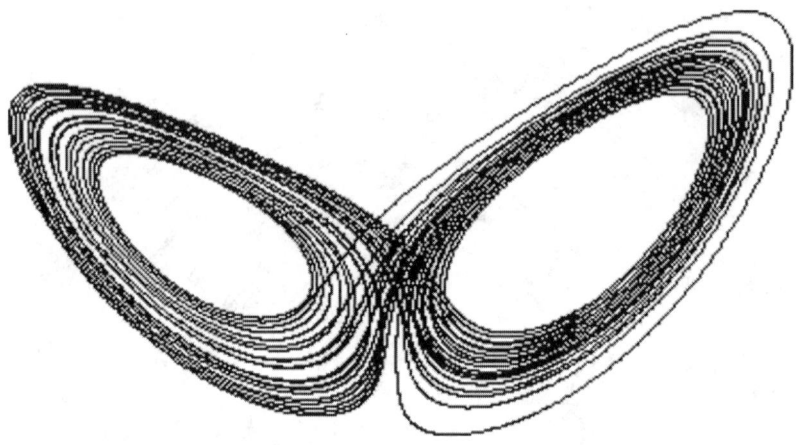

Figure 1.4 A Lorenz or butterfly strange attractor

in a computer simulation, the representation evolves infinitely in the multidimensionality of state space. As we watch the trajectory move through state space, the memory of initial conditions is lost as new information replaces it. Because the finite bounds of the strange attractor must encompass exponentially diverging orbits (a potentially infinite process), a stretching and folding operation takes place, analogous to the stretching and folding that occur as we knead bread dough. If we add a drop of food coloring to the dough and then perform several iterations of the kneading process, we cannot locate the original drop, although we can now see streaks of color diffused throughout; nor can we know the patterns that those streaks will take as we perform further iterations. As a strange attractor evolves over time, the starting place becomes lost: "The stretching and folding operation of a chaotic attractor systematically removes the initial information and replaces it with new information: the stretch makes small-scale uncertainties larger, the fold brings widely separated trajectories together and erases large-scale information."[52] Always in the process of becoming, the strange attractor visually demonstrates our inability to retrodict the past state of a system or to predict its future state. Although we cannot make precise predictions, computer simulation enables us to discern the system's overall behavior over time—to apprehend visually its disorderly order.

Geometrically, the strange attractor exists in a fractal dimension. Whereas traditional geometry studies smooth shapes, such as circles, cones, and cubes, fractal geometry studies rough shapes. The geometrical form of a fractal can occur as a natural object—for example, the shape of trees and coastlines. It can also occur as "a mathematical deduction from an underlying chaotic dynamic."[53] In order to define such shapes mathematically, mathematicians use fractional numbers: for example, a protein molecule has a fractal dimension of 1.7.[54] When discussing one of astronomer Michele Hénon's computer simulations of deterministic chaos, Ekeland notes how the fractal quality "makes possible a continuous transition from the regular, predictable motion of the central trajectories to chaotic, unpredictable motion of the outer trajectories" with "chaotic regions" containing "islands of order."[55] The strange attractor exists in a fractal dimension because, technically, although it is a one-dimensional object (a line), it traverses but does not fill three-dimensional space.

Common to both natural and mathematically deduced fractals is the quality of similarity across scale. In trees, for example, the pattern of the branches repeats on a smaller and smaller scale, and the inlets and promontories of coastlines have their own inlets and promontories. In the strange attractor, similar structures exist at different scales. Whereas the strange attractor provides a useful way of characterizing the bounded randomness of a chaotic system, the fractal does likewise in characterizing its property of similarity across scale. As we shall see, each method illuminates the study of narrative.

Toward a Dynamics of Narrative

In the late 1980s and early 1990s, chaos theory became a pop culture phenomenon, with graphic designs of strange attractors and colorful fractals appearing on tee shirts and posters and with phrases like "the butterfly effect" entering everyday speech. Michael Crichton's 1990 mega-bestseller *Jurassic Park* and the films based on it epitomize the influence of chaos theory on popular culture. Because terms such as "chaos" and "determinism" have philosophical as well as scientific resonance, a great deal of excitement permeated cultural studies about this apparent paradigm shift in scientific thinking. A downside to this excitement was the possibility that chaos theory would be simply a passing fad, and some scholars who initially embraced its possibilities did move on to other theoretical frontiers. In the decade and a half since it burst onto the scene, however, chaos theory has demonstrated that it is not simply a passing fad but a viable and important means for generating new insights within all the disciplines.

Whether chaos theory marks an actual paradigm shift is arguable. After all, classical physics still works very well for explaining many aspects of the observable world. Nonetheless, chaos theory marks a shift in the way we conceive order and disorder, predictability and unpredictability, and deterministic behavior and random behavior. It has had a profound influence on the way in which science is done. Nonlinear dynamics continues to be a significant area of research in the physical sciences, the life sciences, and the social sciences.[56] In the humanities, the rich vein of chaos-theory applications continues to be mined, the insights offered by chaos theory providing us with new understandings of cultural, as well as natural, processes.

The humanist co-optation of chaos theory has often been encouraged by scientists themselves. Prigogine and Stenger's 1984 *Order out of Chaos* has as its subtitle *Man's New Dialogue with Nature,* and the book argues at length for the philosophical implications of chaos theory, seeing it as occasioning a "reenchantment" of the natural world. Crutchfield et al. conclude their groundbreaking 1986 article "Chaos" with a rumination on chaos, creativity, and free will:

> Innate creativity may have an underlying chaotic process that selectively amplifies small fluctuations and molds them into

macroscopic coherent mental states that are experienced as thoughts. In some cases, the thoughts may be decisions, or what are perceived to be the exercise of free will. In this light, chaos provides a mechanism that allows for free will within a world governed by deterministic laws.[57]

In the two decades following these studies, humanists have begun to develop what these scientists merely hinted at, engaging in full-scale explorations of the workings of chaos in artistic production.

Chaos theory has enlivened literary and cultural studies, stimulating an intellectual boundary crossing (at least from the humanist side) perhaps not seen since the two-culture rift arose out of the triumph of classical physics. With regard to literary applications of chaos theory, scholars and creative writers have taken many different approaches. Chaos has been explicitly thematized in fictional texts, such as Tom Stoppard's intellectual tour de force *Arcadia*, Darren Aronofsky's mind-boggling film *Pi*, and the *Jurassic Park* phenomenon.[58] Scholars of chaos such as Hayles, Weissert, and Alexander Argyros have reevaluated post-structuralist tenets in light of chaos theory, and Hayles in particular has discussed at length the cultural implications of deterministic chaos.[59] Colin Martindale has even attempted to model literary history by way of chaos equations.[60] Specific texts or sets of texts, from *Wuthering Heights* to *The Bluest Eye*, have been reexamined using principles from chaos theory.[61]

The crossing of the disciplinary boundaries into chaos-science territory has invited controversy, however. Paul Gross and Norman Levitt, for example, claim that the study of science requires time and labor incompatible with a humanist's training, and they speak cuttingly of the humanist who would venture into the dynamicist's realm: "Thus we encounter . . . essays that make knowing reference to chaos theory, from writers who could not recognize, much less solve, a first-order linear differential equation."[62] Perhaps the most vexed episode in the so-called science wars has been the notorious "*Social Text* Affair," whereby physicist Alan Sokal wrote a parodic essay linking "postmodern" science and postmodern theory and passed it off as sincere.[63] In subsequent publications, Sokal has continued to take to task literary theorists who draw upon contemporary science for what he claims are "meaningless or absurd statements, name-dropping, and the display of false erudition."[64]

Granted, some literary theorists have used scientific terminology imprecisely and made unsupportable claims. But, as Theodore Braun and John McCarthy point out, the "trade-off" that takes place "between rigor and vigor" is positive overall: "Whoever ventures out from one's own (disciplinary) language into another brings to that endeavor (and world view) a perspective the 'native speaker' lacks. An enrichment takes place, which, in turn, dilutes the rigor of the original conceptual system by introducing 'foreign elements' into it."[65] Literary theorists are not claiming to be scientists; they are instead laying claim to a conceptual trove that enhances literary studies—and enhances science as well.

Sokal and Jean Bricmont complain, "[W]e fail to see the advantage of invoking, even metaphorically, scientific concepts that one oneself understands only shakily when addressing a readership composed almost entirely of non-scientists."[66] The advantage is two-fold, however. First, by drawing on scientific concepts literary theorists find new ways of articulating and apprehending the complex workings of a literary text. After all, new concepts lead to new ways of seeing. In the physical sciences, for example, the concept of the fractal has taught us to discern the disorderly organization of the coastline or the self-similarity in tree branches. Similarly, chaos theory can enhance our understanding of the dynamics of literary texts because it enables us to see what we have not seen before. Second, and even perhaps more important, the drawing on scientific concepts encourages literary theorists and their nonscientist readership to engage with science, to return to studies that they were forced to abandon as they pursued specialized degrees and to realize that science is a much a part of culture as art. We thus rediscover a valuable source of knowledge.[67] Although most of the boundary crossing has involved literary theorists venturing into scientific terrain, I believe that as scientists embark on the opposite journey, both disciplines will gain.

My own approach to chaos theory is to demonstrate how certain narrative structures resemble chaotic nonlinear dynamical systems. Before turning to narrative chaotics, I want to address the issue of narrative as a dynamical system in general. In a 1996 *Lingua Franca* article, Steven Johnson claims that "dressing up the old-fashioned values of good storytelling in the new language of complexity does have a certain emperor's-new-clothes air about it."[68] He cavils at what he sees as the "dangerous game" of regarding a text as a complex system:

> The greatest problem with literary chaotics may be that a work of literature is not a system at all, in the Santa Fe sense of the term—that is, a dynamic mix of agents interacting in real time. Novels, for example, may be *about* complex systems (cities, economies, ecosystems, and so on) and they are certainly the *products* of complex systems (the neural nets of the human mind), but they themselves are language-based, static, dictated from the outside.... Novels do *not* self-organize—that's why we need novelists.[69]

Johnson's point about self-organization is somewhat disingenuous. Certainly, novels do not self-organize so that letters thrown together spontaneously emerge by themselves into meaning, like sentient Scrabble tiles. But Johnson ignores how a writer's organization of letters, words, motifs, and episodes is always subject to reorganization as readers engage in the reading process. For Johnson, the literary text is "static," fixed in its meaning. It is simply a product and, as such, it cannot be a system.

Yet readers engaged in the reading process do indeed encounter "a dynamic mix of agents interacting in real time." Granted, the structure of a traditional narrative is static in that readers deal with a printed text "dictated from the outside," but if we replace the word "structure" with Paul Ricoeur's more apt "structuration," the dynamics of narrative emerges. As Ricoeur points out while discussing Aristotle's *Poetics*, "Structuration is an oriented activity that is only completed in the mind of the spectator or reader."[70] Narrative structuration is an ongoing dynamical process, not a fixed product, as Johnson envisions it.

Ricoeur's description of the "dynamic of emplotment" is pertinent. In narrative, "the textual configuration mediates between the prefiguration of the practical field and its refiguration through the reception of the work."[71] In other words, "the prefiguration of the practical field" becomes the story events, "the textual configuration," the shaping of those events into the narrative plot, and "the reception," the reader's act of making meaning of that plot.[72]

Similarly, dynamicists engage in a "dynamic of emplotment." They observe the empirical data of a complex system—for example, atmospheric variables or the mass and rate of speed of drops falling from a dripping faucet—that they then build into a set of equations or model. Using Ricoeur's terminology, we would call this model the prefiguration of the practical field. Computer simulation then uncovers a geometry of the system's behavior over time in the state-space portrait that emerges—what we might, following Ricoeur, call the textual configuration. This graphical display for analysis compares with the narrative put forward for the reader's reception. Weissert makes an explicit connection between the work of dynamicists and the work of those who construct narratives: "Modeling a physical system is nothing more or less than fixing the constraints for narrative construction, when we conceive of that system as a sequence of events."[73] In essence, dynamicists read and interpret patterns. Whether we confront a literary text or a dynamical system, similar processes of meaning making are concerned—processes that involve a complex interaction of system, model, and modeler.

The figure of the strange attractor—that simulation of a system's behavior over time—drives home the way in which the object of study is integrally connected to the meaning-making activity of the subject who studies it. Fraser wisely observes, "One often sees bicycles whose motion could be described as an attractor in phase space, but one never sees those attractors traced out around the bicycles in multidimensional space."[74] That is to say, the attractor does not exist out there in reality, but it is a particular means for describing the behavior of a system. It is a *figure* (in the sense of a trope) of the system's dynamical behavior, created by the dynamicist's manipulation of parameters in state space.

Although my work explores the analogous behaviors of a dynamical system's simulation and a narrative, I do want to point out that there are a few crucial distinctions. Whereas dynamicists alone observe the empirical data, configure the system's behavior, and interpret the resulting simulation, these tasks, in a narrative system, are divided between writers and readers. Writers observe the data and formulate a plot; readers interpret the textual configuration that has been devised by writers. My statement that writers observe the data may in itself be somewhat misleading. With regard to fictional narrative, Gérard Genette notes that we see "the narrative act initiating (inventing) *both* the story and its narrative, which are then completely indissociable."[75] Laurence Sterne, for example, invents the life of the imaginary character Tristram Shandy, the key events of whose life Sterne can then jumble chronologically in the narrative that he writes. Perhaps even more to the point, Peter Brooks notes that "the story is, after all, a construction made by the reader . . . from the implications of the narrative discourse, which is all he ever knows"—a statement that reinforces my point about division of tasks between reader and writer.[76] Writers of narrative fictions draw upon the empirical data of the real world to a greater or lesser extent, but the facts of these fictions are necessarily inventions that the reader regards as the real basis for the plot. Nevertheless, we can say that the writer, like the dynamicist, works with or upon the materials of the real world.

A related qualification concerns the constraints placed on interpretation. The empirical data that dynamicists observe, plot, and interpret are, in fact, fixed. Iterating Lorenz's nonlinear differential equations for cellular convection will result time and time again in a butterfly attractor, as opposed to a funnel attractor. By starting with a different set of initial conditions, dynamicists may set the trajectory going at a different place in the attractor's evolution, but the emergent pattern will nonetheless always reveal the particular attractor to which a particular dynamical system tends within the parameters. These parameters effectually constitute an interpretation of the data. Dynamicists may change them: for example, in the dripping faucet simulation, they may vary the parameters in the equations for elasticity, viscosity, or force of gravity. In doing so, they provide new interpretations in order to find that which most closely matches the physical system that they are modeling.

Readers also vary "parameters" (that is, what they look for in a text) in order to find what they consider the most accurate interpretation. However, because of the ambiguity in language and the many contexts from which texts come and from which readers operate, what one reader gets from a narrative may be very different from what another reader gets—and neither of these readings may have to do with what a writer thinks he or she has written. In a dynamical system, rigid, quantifiable constraints exist. In a narrative system, there are constraints on the emergent interpretive pattern as well: for example, the historical context of the text and of its readers, conventional semantic meanings, conventions of reading, prior readings. However, because so many variables are involved and because these variables cannot be quantified, a narrative system has many more degrees of freedom than has a dynamical system, which accounts for the variety of interpretations that results. Despite these distinctions, a similar process of meaning generation and meaning making pertains to dynamical systems and narrative structuration.

Although the analogy between a narrative and a dynamical system has potential for further exploration, my purpose here is not to so much to deal with narrative structuration in general but a particular kind that I characterize as chaotic. Again, chaotic dynamical systems undermine classical notions of stability, repeatability, predictability, causality, absolute time, and observer objectivity. So do chaotic narratives, manipulating the ordering, repetition, and duration of events and highlighting the interconnection between text and reader.[77]

Specifically, with regard to order, a chaotic narrative may scramble chronology. Rather than the story events being ordered linearly, as if on a time line, past, present, and sometimes even future events intermingle. This device calls into question clear causal connections, for we find that, as we process information, a chronologically later story event has an impact upon an earlier one. Furthermore, we often have difficulty discerning the initial conditions that gave rise to the current situation.

With regard to repetition, a chaotic narrative may put forward several versions of the same event. Because a sensitive dependence on initial conditions determines the evolution of a narrative trajectory, new and different information is imparted with each iteration, a word whose general and math-specific meanings I invoke intentionally. In addition to presenting one event multiple times, a chaotic narrative may also synthesize many events into one, a narrative technique that Genette felicitously terms the "iterative."[78] Each of these devices calls into question a fixed truth of events.

With regard to duration, a chaotic narrative exacerbates the compression and expansion of various episodes. It deliberately draws our attention to the relations that exist between the duration of the events, the duration of the reading, and even the duration of the writing. In doing so, the narrative undermines a notion of absolute time, demonstrating that time differs perceptually at different local levels.

With regard to the interaction between text and reader, a chaotic text may deliberately highlight the dynamics of the reading process. Often, it demonstrates these dynamics within the text itself, featuring a metanarrative and characters engaged in interpretation—or misinterpretation—of it. A chaotic narrative may commit metaleptic transgressions by crossing the boundary between narrative level: for example, between the world of the text and the world of the reader, as when "author" Tristram surmises "his" reader's thoughts. Metanarrative structure and metaleptic transgressions illuminate how a text's meaning emerges from the interaction of text and reader.

The four narratives that I examine in the following pages foreground such a variety of chaotic elements that their overall structures seem to fit John Casti's succinct description of a strange attractor—"one big tangled mess."[79] Granted, previous critics have disentangled these "tangled messes"; the act of interpretation has proceeded apace. But, by taking into account the conceptual shift that chaos theory has engendered in our thinking, we can look at these texts in new ways, understanding them in terms of a bounded randomness, infinitely evolving within certain constraints. Doing so enables us to go beyond the either/or debates that so often occur—between spatial and temporal models, between what is there in the text and what the reader brings to it, between fixed meaning and undecidability, between form and content.

First, as we examine chaotic narratives through the theoretical lens of chaos, we travel beyond a simple spatial-versus-temporal dichotomy. We are made aware of the dynamical nature of textual heuristics. The trajectories on the strange attractor never reach a fixed point; the system is always in the process of becoming. Nevertheless, a pattern emerges—and will continue to do so. The strange attractor is simultaneously a spatial configuration and a temporal continuum—a spatialization of temporal process and a temporalization of spatial form. It is thus an apt figure for our interaction with the chaotic narrative—a determinate, fixed spatial product when we consider the words on the page and an indeterminate, ongoing temporal process when we consider the meaning that we derive from those words. This dichotomy foregrounds the reader's involvement in the meaning-making process, analogous to the dynamicist's involvement in generating meaning from the strange attractor, which leads me to my next point.

Second, the chaotic text pointedly invites our participation and acknowledges our involvement in the meaning we derive from it. It compels the writerly reading that Roland Barthes idealizes:

> The writerly text is a perpetual present, upon which no *consequent* language (which would inevitably make it past) can be superimposed; the writerly text is *ourselves writing*, before the infinite play of the world (the world as function) is traversed, intersected, stopped, plasticized by some singular system (Ideology, Genus, Criticism) which reduces the plurality of entrances, the opening of networks, the infinity of languages.[80]

I would add the caveat that these "entrances" lead to the basin of attraction. There may be an infinite number of interpretations of a particular text, but they all fall within a bounded area.

Third, the chaos theory lens enables us to appreciate the infinite play of signification within a bounded arena of truth or meaning—a basin of attraction. In the case of a multiperspectival narrative, for example, our tendency has either been to regard it as leading to some fixed "truth" of events or as pointing to the relativism of all truth. By considering the various narrative trajectories as lying on a strange attractor, we can think in terms of a pattern that approximates a truth that can never be achieved. Chaos theory also enables us to reevaluate iterative sequences in texts; such sequences represent the unrepresentable, what does not exist, yet they approximate a truth common to all the events that they synthesize into one. Chaos theory reconciles the opposing notions that the chaotic narrative advances a closed, fixed, determinable meaning and that its meaning dissipates into indeterminism.[81] Linda Alcoff points out that "the notion that all texts are undecidable cannot be useful for feminists." As she suggests, affirming a text's undecidability precludes critique of its ideological import. I would suggest that a chaos-theory reading enables us to affirm a text's undecidability *and* offer a critique of its ideological import.

Perhaps most important for the particular texts I examine, a chaos-theory reading helps us make a connection between form and content. A common complaint against structuralist narratology is that, because of its focus on form, it ignores the social conditions out of which narratives arise and which they reflect. In his reevaluation of *Narrative Discourse*, Genette postulates two poles: "[W]e can no doubt observe that literary studies today oscillate between the philately of interpretive criticism and the mechanics of narratology."[82] The formalism of structuralist narratology often opposes the interpretive criticism practiced by post-structuralist theorists doing work in feminism, postcolonialism, queer theory, or racial and ethnic studies. For example, Susan Sniader Lanser notes with regard to the polarization between narratology and feminism: "With a few exceptions, feminist criticism does not ordinarily consider the technical aspects of narration and narrative poetics does not ordinarily consider the social properties and political implications of narrative voice."[83] Recent work in narratology has increasingly attempted to bridge the form-content gap, and a chaos-theory reading provides us with additional tools for doing so.

The significance of the strange-attractor structure is that it *necessarily* connects form with content. Again, a strange attractor occurs when the attracting point in a chaotic dynamical system has become unstable, thus concurrently attracting and repelling the system trajectory. There is an actual attracting point (or points) in the system. In the narrative text, the attracting point comprises motifs that concurrently attract and repel the writer. In the four texts that I examine, such motifs include the death of the text, the death of the self, erotic and aesthetic fulfillment, precise memory, determinate identity, gender norms, and the racial conflicts of the American South. To examine these texts' structures through a chaos-theory lens requires that we be aware of the content that prompted the chaotic structure. And, because the content in each of the four narratives is different, the strange attractors function differently as well, just as they do in different physical systems.

Franco Moretti reminds us that "the social aspect of literature resides *in its form*" (his emphasis).[84] The structuration of a chaotic narrative is different from the structuration of other narratives—and, more to the point, it is different for a particular purpose. When a narrative plays games with repetition, order, and duration, and when a narrative pointedly implicates the reader in the process of meaning making, these devices become integral to the meaning itself. In the following pages, I demonstrate that the chaotic narratives I have selected emerge from concrete issues with which the writers attempt to deal. Their disorderly order serves as a deliberate resistance to the cultural determinations of their time and to the interpretive acts that would fix them.

CHAPTER 2

Narrating against the Clockwork Hegemony: *Tristram Shandy*'s Games with Temporality

> *In a word, my work is digressive, and it is progressive too,—and at the same time.*
>
> —*Tristram Shandy*

> *The Plan, as you will perceive, is a most extensive one,—taking in, not only, the Weak part of the Sciences, in which the true point of Ridicule lies—but every Thing else, which I find Laugh-at-able in my way—.*
>
> — Laurence Sterne, *Letters*

> *Time's out of rule; no clock is now wound-up: TRISTRAM the lewd has knock'd Clock-making up.*
>
> —*The Clockmakers Outcry Against the Author of* The Life and Opinions of Tristram Shandy

Laurence Sterne's *The Life and Opinions of Tristram Shandy* begins not simply with the hero's birth, but (apparently) his conception. Significantly, because of the Shandy family's proclivity for a Lockean association of ideas, the moment of conception is connected with the winding of the clock, as if to suggest that Tristram, or the "homonculus" that will become Tristram, has been born into clock time. But we must bear in mind that Walter Shandy may have forgotten to wind the clock, as Mrs. Shandy's question ("have you not forgot to wind up the clock") suggests.[1] This circumstance is in keeping with the fact that Tristram and his text are subject to something other than, in Stephen Kellert's felicitous phrase, "the clockwork hegemony" instigated by Newtonian science.[2]

In many ways, Tristram is the chaotic text par excellence, with regard to both temporal order and temporal duration. Its jumbling of chronology constitutes a species of disorderly order. Unlike a plot that moves inevitably, predictably to its own cessation, as plots do in the linear novels common to the eighteenth century, *Tristram Shandy*'s plot points suggestively to a potentially infinite evolution, analogous to that of the strange attractor.[3] Narrator Tristram's self-conscious running commentary on the disorderly narrative trajectory makes clear that Sterne deliberately works against the deterministic tendencies of the linear plot—and the deterministic thinking endemic to the eighteenth century. Furthermore, the text entangles six "levels" of temporal duration, three levels within the fictional world of the text that point to three levels existing in an extra-textual temporal reality. Whereas critics have tended to discuss the text in terms of the opposition it draws up between objective clock time and the subjective time of consciousness, this entanglement of various levels, accompanied by narrator Tristram's self-reflexive commentary, calls the very notion of an objective or absolute time into question, enabling us to understand the limitations of the clockwork hegemony for explaining the chaotic operation of time in our lives.

In examining *Tristram Shandy* through the insights provided by chaos theory, I do not mean to suggest that Sterne had some sort of incipient understanding of this quintessentially postmodern science. Rather than the disorderly order or deterministic chaos that it has come to mean in our current scientific paradigm, chaos for Sterne and his contemporaries would have meant an absence of any order, along the lines of Milton's description of Chaos's realm in *Paradise Lost*: "Rumour next and Chance, / And Tumult and Confusion all embroil'd, / And Discord with a thousand various mouths" (2.965–67).[4] The realm of chaos would have been a frightening one to an eighteenth-century writer such as Sterne. Nevertheless, he seems to have sensed that the ordered universe envisioned by Newtonian scientists was not amenable to him. The narrative dynamics of *Tristram Shandy* deliberately challenge the reigning scientific paradigm of Sterne's time—the determinism of Newtonian science, to which the great linear narratives of the mid-eighteenth century conform. Although Sterne lacks the means for articulating a science of chaos, his text foregrounds the disorderly order and the complexity of time's flow that Newtonian science occludes.[5]

The Strange Attractor of Death

Some of the most evocative images in *Tristram Shandy* are graphical—the black page, yawning like an open grave that follows the announcement of Yorick's death; the "flourish" Trim makes with his stick to represent an unmarried man's freedom; the five lines, interrupted by zigzags and curlicues that Tristram tells us, represent the narrative movement of the first five volumes; the marbled page, "motley emblem of my work" (226); and so forth. With the significant exception of the black page, all these images are dynamical, and we can imagine that, had Laurence Sterne had access to the graphical representations of contemporary dynamicists, he would have included an image of a strange attractor as an appropriate visual representation of his text. For *Tristram Shandy* exemplifies a chaotic dynamical system, a bounded arena of infinite possibility. Its nonlinear structure is a reaction to the grand linear narratives of the eighteenth century, whose trajectories move predictably to a steady state where their action ceases. The narrative trajectory of *Tristram Shandy* hovers in a simulacrum of perpetual motion over the powerful, but unattainable, attracting points of sex and death, bearing strong similarities to the Lorenz or butterfly attractor with its two attracting points.

What Stephen Kellert labels the "linear prejudice" of classical physics applies as well to eighteenth-century novels, which move forward with deterministic inevitability to their culmination in marriage or death.[6] Let us take Henry Fielding's *Tom Jones* as a paradigm eighteenth-century novel. It starts with Jones's birth, advances predictably (even the twists in the plot are traceable in retrospect), and ends with both his marriage and death—the death of the imprudent Jonesian self. Granted, the text is rife with digressions and interpolations, but we can regard these as perturbations or pockets of chaos that are smoothed out by the dominance of the overall linear tendency. Although we cannot predict precisely where Fielding will take us next, we can nonetheless make fairly accurate predictions that, if Tom finds Sophia's muff in one chapter, he will get it back to her in a subsequent one, and that his incarceration will be followed by his release. We know that, in Roland Barthes's terms, the hermeneutic sentence upon which the narrative movement is predicated will be answered.[7] In other words, the mystery of Tom's birth will be solved. After finishing the novel, we can go back and discern the strict causality connecting all the sequences, as Ronald S. Crane's influential essay "The Plot of Tom Jones" elegantly demonstrates.[8]

The linear narrative and predictability would seem to go hand in hand. Although his focus is not linear narrative per se, Barthes sees an affinity between narrative and deterministic thinking, an opinion no doubt reinforced by his experience of the pronounced linearity of that most popular of modern narrative forms, the novel: "Everything suggests, indeed, that the mainspring of narrative is precisely the confusion of consecution and consequence, what comes *after* being read in narrative as what is *caused by*, in which case narrative would be a systematic application of the logical fallacy denounced by Scholasticism in the formula *post hoc ergo propter hoc*."[9] This linear prejudice reaches its apotheosis in the great autobiographical novels of the nineteenth century, whose action begins with the self's fall into linear time and ends with the "death" of the narrated self into the narrating self, whereupon all seemingly casual events are gathered up into the overarching causal pattern and the narrating self "writes" from the atemporal state of the narrating instance. Epistolary novels that retrace prior events from another perspective (e.g., *Clarissa*) are sometimes locally nonlinear; however, they are globally linear, moving straightforwardly along the time line.

Although Sterne's text delighted readers from the outset, it nevertheless fell victim to linear prejudice, confounding the expectations of contemporary reviewers. William Kenrick, for example, advised Sterne to pay "a little more regard to going straight forward," concerned (justifiably, we find) lest Tristram give the readers "the slip in good earnest, and leave the work before his story be finished," and Edmund Burke complained that Sterne's "digressions . . . instead of relieving the reader, become at length tiresome" and that the book itself "is a perpetual series of disappointments"—in essence, disappointments to a deterministic reader.[10] The text's full title—*The Life and Opinions of Tristram Shandy, Gentleman*—does indeed set up expectations of a birth-to-death linear structure, although the term "Opinions" should alert us to perturbations in it.

Certainly, *Tristram Shandy* is bound, like all autobiographies, by two definitive events, Tristram's birth and the death of the narrated self. It begins, in fact, with the quintessential beginning, describing not simply the birth, but what appears to be the actual conception of the protagonist. Tristram the narrator congratulates himself for the feat: "[R]ight glad I am, that I have begun the history of myself in the way I have done; and that I am able to go on tracing every thing in it, as *Horace* says, *ab Ovo*" (7). Tristram aims to reach the point where the narrated self becomes the narrating self: "[W]hipp'd and driven to the last pinch, at the worst I shall have one day the start of my pen" (286). But the words preceding this statement accurately prophesy that the aim will never be realized: "I shall never overtake myself" (286). Tristram cannot, ultimately, discover the initial conditions that gave rise to his narrative trajectory or achieve narrative death, except, of course, in the sense that the text does and must physically come to an end.

Recall that when we map the behavior of a pendulum, it exhibits classical deterministic behavior, and it falls onto a fixed-point attractor—that is, an attractor that shows the system's movement toward eventual cessation. We might say that in the linear (auto)biographical novel, the birth of the protagonist initiates a narrative trajectory that is attracted to and comes to rest upon the fixed point of the protagonist's death, whether symbolic or real. In *Tristram Shandy*, however, the two unstable attracting points of sex and death ensure that the narrative trajectory seemingly never comes to rest, giving us a sense of infinite potentiality. As Robert Alter fittingly points out, the text attempts to "make us repeatedly aware of the infinite horizon of the imagination" within "a finite narrative form."[11] It works against our being able to retrodict a prior or predict a future state of the system, just as occurs with a strange attractor.[12]

In *Tristram Shandy*, Sterne emphasizes the impossibility of pinpointing Tristram's initial conditions. He gives us reason to suspect that the interrupted intercourse of the Shandys may not be the actual moment of conception—that Tristram may, in fact, be illegitimate.[13] Significantly, legitimation is a potent signifier of deterministic causality, demonstrating a clear connection to an origin and guaranteeing a (patri)lineal outcome. Tom Jones, is, of course, illegitimate, but the plot in which he is inscribed is predicated upon the discovery of his origin—a discovery that seemingly clarifies all prior events and brings him back, albeit obliquely, to the patriarch, Squire Allworthy.[14] Whether Tristram is indeed illegitimate is indeterminate and indeterminable. Sterne avoids the clarification that would shed light on prior events.

Even if Tristram's poor homunculus were brought into being during the ill-fated coupling, Tristram himself discovers that he must go further back than the conception, to account for himself. But when he does, he confronts a tangle of rapidly proliferating choices: "To sum up all; there are archives at every stage to be look'd into, and rolls, records, documents, and endless genealogies, which justice ever and anon calls him [a man] back to stay the reading of:— In short, there is no end of it" (37). Tristram does not actually supply us with these "endless genealogies," but we are encouraged to picture a multiplicity of narrative strands moving backwards. For Tristram, as well as the reader, initial information is replaced with new information, as with a strange attractor, and he cannot locate a fixed origin that would account for who and what he is.[15]

Nor can Tristram or we predict the future course of his narrative trajectory. Whereas within a classical dynamical system, events can be predicted, within a chaotic dynamical system, events are unpredictable beyond a certain point in time. In *Tristram Shandy*, we find certain discrete linear sequences, generally fully contained within one of the brief chapters— Dr. Slop's reading of Ernulphus's curse; the hot chestnut falling into the "hiatus in *Phutatorious's* breeches" (321), followed by the unfortunate result; Toby and Trim's march from the bottom of the avenue to Widow Wadman's door; and so forth. Without such proairetic sequences, the text would be incomprehensible. But the sequences constitute part of longer sequences, and these rarely, if ever, proceed linearly. Although in his edition of *Tristram Shandy*, James Work points out that "the leading overt actions of the story, developed through two overlapping sequences [Walter Shandy's household affairs and Uncle Toby's courtship], are arranged within each sequence in perfect chronological order" (xlviii), these sequences are still in process when we have reached the end of the text, and endless possibilities seemingly exist for the narrative trajectory to revisit these areas of the text's "state space." We may make a global prediction about Tristram's future—that he would eventually reach the point where he sets out to write his *Life* (a situation that never actually occurs in the text). We cannot, however, make any sort of prediction about where the narrative trajectory will be by the time we turn the page.

Certainly, one can argue that the text has a predictable quality. We are not surprised that, when Toby discerns "the transverse ziz-zaggery" of Walter's approach to his coat pocket (160–61), he will be reminded of the battle of Namur, or that, when he hears that Dr. Slop is in the kitchen making a bridge, he thinks of his destroyed drawbridge. We expect Walter to come up with quirky arguments on esoteric matters, and we expect our narrator to give the most innocent subjects a risqué turn. But predictability of character is not the same as predictability of sequence. Even if Sterne structures the sequence of episodes so as to represent the path followed by the mind as it associates ideas, the path does not unroll linearly and inevitably. The mind jumbles temporal order, connects like with unlike. Mental events have causes, to be sure, but, as in a chaotic dynamical system, we cannot determine which and how many causes lead to a particular effect and which and how many effects derive from a particular cause. Indeed, the text serves as an implicit demonstration of the fact that linear causality inadequately models the complex workings of the mind, reminding us that the great explanative narrative put forward under the Newtonian paradigm leaves the human element out of the equation.

The text is insistently nonlinear. In the fifth chapter of the first volume, Tristram is born, and Sterne gives us an exact date, as if to suggest that we will be proceeding according to a strict chronological order. As A. A. Mendilow has so effectively demonstrated, Sterne "astonishes us by the accuracy with which the dates . . . are nevertheless made to cohere."[16] But after telling us of his birth, Tristram goes backwards to provide us with the history of the midwife, gives us the dedication, which should indeed be outside the narrative proper (if there can be such a thing), recounts Yorick's history, and then jumps ahead to Yorick's death—an account followed by black pages, a negative image of the pages that customarily follow the end of a text. In the ninth volume, chapters 18 and 19 follow chapter 25, and their proper place is filled by white pages, falsely representing the climax (the text's, Toby's) that cannot occur. In one volume, we may be traipsing around Europe with Tristram the narrator, the narrating instance itself evolving in time; in the next, we may be privy to the emotional modulations of Toby in love. Each of these sequences is itself riddled with interpolations and temporal jumps. With regard to the text's temporal (dis)ordering, we thus find similarity across scale, that characteristic of dynamical systems whereby structural similarities occur at both global and local levels. Such scaling is often noted in fractal forms, wherein the patterns we discern at one level replicate themselves on smaller and smaller scales ad infinitum. Ultimately, at all levels, we have no way of knowing when the narrative trajectory will jump to another part of the attractor basin, such as the trajectory of the strange attractor Edward Lorenz discovered when he iterated equations for convection: "it crosses from one spiral to the other at irregular intervals."[17]

Unsurprisingly, the unstable attracting points between which the narrative trajectory jumps are death and sex—the former the great unknowable and the latter the great unmentionable. These attracting points, powerful draws for Sterne and his culture, concurrently and continually attract and repel the narrative trajectory.[18] Sterne—that womanizing clergyman racked with consumption—is fascinated by the generative act he must not explicitly describe and the definitive act that he cannot. Of relevance here is Peter Brooks's description of narrative as "the thrust of a desire that never can quite speak its name—never can quite come to the point—but that insists on speaking over and over again its movement toward that name."[19] According to the constraints under which he works, Sterne can represent the attracting points of sex and death only through structural deferral and figural displacement.

Although, by jumping between these two attracting points, the narrative trajectory of Tristram Shandy resembles a Lorenz attractor, there is a significant difference. A rigid global predictability governs the trajectory of the Lorenz attractor, for it jumps between attracting points in a strictly alternating sequence. Consider, for example, a waterwheel, a dynamical system that can manifest a strange attractor. The spin of a waterwheel can become chaotic as water flow increases, as James Gleick describes:

> As buckets pass under the flowing water, how much they fill depends on the speed of spin. If the wheel is spinning rapidly, the buckets have little time to fill up.... Also, if the wheel is spinning rapidly, buckets can start up the other side before they have time to empty. As a result, heavy buckets on the side moving upward can cause the spin to slow down and then reverse.[20]

When we map the waterwheel's behavior in state space, we end up with a Lorenz or butterfly attractor; the trajectory's jump from one "wing" to the other represents the unpredictable reversals of motion that the waterwheel undergoes.

The text's narrative trajectory, however, does not so much alternate between sex and death as move toward, then away from, climaxes. The attracting points are, in fact, integrally related, for each marks the culmination of an apparently linear sequence—life or love. When Slawkenbergius gives his critical disquisition on the movement of plot toward its culmination, he may as well be speaking of sexual, as well as narrative, climax (an appropriate ambiguity considering that the story he tells has to do with the Strasburgers' feverish desire to satisfy themselves by touching Diego's huge "nose"): "The *Epistasis*, wherein the action is more fully entered upon and heightened, till it arrives at its state or height called the *Catastasis* . . . or the ripening of the incidents for their bursting forth in the fifth act" (266). Interestingly, the Strasburgers' situation is anticlimactic, for Diego never returns to satisfy their desires. And, although the Strasburgers' anticlimax is the climax of Slawkenbergius's story, we too are left unsatisfied, never finding out if the "nose" itself is real. We never get to the thing itself, for, as we all know, despite Tristram's asseverations that "by that word I mean a Nose, and nothing more, or less" (218), a nose does not mean a nose, any more than sausages means sausages and buttonholes mean buttonholes.[21] Slawkenbergius's tale epitomizes Sterne's procedure of structural deferral and figural displacement.[22] We should bear in mind that the strange attractor is both temporal continuum and spatial configuration. Similarly, it is the entanglement of the temporal (structural deferral) and the spatial (figural displacement) that creates the meaning structure of *Tristram Shandy*.

It is a given that, just as Tristram attempts to escape death through his wild zigzag across Europe, *Tristram Shandy* attempts to avoid its own "death" through its temporal disorder.[23] The text resists an ending that would be a result of its prior state and that would enable us to see an overarching causal pattern. The fact that Sterne's actual death left the text cut off in the middle of Toby's amours and the middle of the story of the Cock and Bull is beside the point, for Sterne had been aiming for a text-in-process all along. We might even say that Sterne's actual death facilitated the text's avoidance of its own.[24] We are in a state of endless deferral. That "perpetual series of disappointments" that Burke deplored is essential to Sterne's project, and we notice that the closer any sequence comes to reaching a climax, the more interruptions and temporal leaps occur. Digressions indeed "are the life, the soul of reading," as Tristram exclaims, for they keep "the whole machine . . . agoing" (73)—keep the text from reaching its end.

In a felicitous instance of similarity across scale, Trim's "The Story of the king of Bohemia and his seven castles" replicates in miniature the overall narrative movement of the text itself. Trim's story never gets beyond the title (set off in the text four times, the last three with the inaccurate addition "continued") and the first incomplete sentence, and it ends up being displaced by Toby's history of gunpowder and Trim's history of his amours with the Fair Beguine, whose climax (literal and figurative) is interrupted by Toby's unwittingly periphrastic comment about what Trim must have done once his "passion rose to the highest pitch"—that is, clap the Beguine's hand to his lips and make a speech (375). When Toby later asks what became of the story, Trim replies, "We lost it, an' please your honour, somehow betwixt us," (381) and we may well feel that the story of Tristram has itself been lost, just as William Kenrick feared.

The fact that the narrating instance is itself subject to time's movement ensures that the death of Tristram's narrated self can never occur. Tristram's precise dating of when he is writing indicates that his history advances as he purportedly writes, chronology being used here to subvert "the clockwork hegemony" instigated by Newtonian science. That situation, of course, leads to the famous Shandean paradox, whereby the longer Tristram is at his writing, the further behind he gets:

> I am this month one whole year older than I was this time twelve-month; and having got, as you perceive, almost into the middle of my fourth volume—and no farther than to my first day's life—'tis demonstrative that I have three hundred and sixty-four days more days of life to write just now, than when I first set out; so that instead of advancing, as a common writer, in my work with what I have been doing at it, I am just thrown so many volumes back.... It must follow, an' please your worships that the more I write, the more I shall have to write. (286)

For Tristram, writing does not move inevitably to its own cessation, but it is endlessly generative as it moves toward the climax it must never reach.

The apparent impotence of the Shandy family serves as an appropriate figure for the climaxless text. Indeed, the text is climaxless in more than one sense. As in the story of Tristram's "conception" and that of the Fair Beguine, Sterne temporally defers and figuratively displaces the sexual climaxes themselves. *Tristram Shandy* speaks endlessly around sex but never directly of it. As we progress through the narrative, we acquire more and more means by talking around the subject, with each iteration of a particular motif—such as noses and sausages—enabling the strange attractor's evolution in the state space of the text.

Whereas the narrative trajectory of *Tristram Shandy* challenges the linear prejudice of Sterne's age, the commentary in the text itself challenges the deterministic predictability of Newtonian science. Tristram makes a pseudosolemn prediction of scientific progress achieving a sort of Laplacian vantage:

> Thus,—thus my fellow labourers and associates in this great harvest of our learning, now ripening before our eyes; thus it is, by slow steps of casual increase, that our knowledge physical, metaphysical, physiological, polemical, nautical, mathematical, ænigmatical, technical, biographical, romantical, chemical, and obstetrical, with fifty other branches of it, (most of 'em ending, as these do, in *ical*) have, for these two last centuries and more, gradually been creeping upwards towards that $A\kappa\mu\eta$ of their perfections, from which, if we may form a conjecture from the advances of these last seven years, we cannot possibly be far off. (64)

Significantly, the perfection of knowledge "will put an end to all kind of writings whatsoever" (64)—an achievement against which *Tristram Shandy* directly and forcefully testifies.

In the person of Walter Shandy, Sterne mocks the progressive impetus of Newtonian science. Walter, the ultimate systematizer, attempts to weigh all the variables of situations to determine the future course of his offspring. The culmination of his systematizing is the TRISTRA-*poedia*, which, as Tristram tells us, is intended "to form an INSTITUTE for the government of my childhood and adolescence" (372). Little but crucial circumstances, however, overturn all of Walter's best-laid plans: for example, the fit of wind that leads to Mrs. Shandy lying-in at Shandy Hall, the unfortunate phonetic similarity between "Trismegisthus" and "Tristram," and Trim's appropriation of the leaden weights from the sash pulleys of the nursery windows, which leads to Tristram's involuntary circumcision—or worse. Walter's scientific optimism is belied by his experience, which consistently demonstrates a sensitive dependence on initial conditions, wherein little causes, amplified by feedback, give birth to great and unpredictable effects. Significantly, Walter himself argues against the classical tendency to disregard small uncertainties in a system:

> Knowledge, like matter, he would affirm, was divisible *in infinitum*;—that the grains and scruples were as much a part of it, as the gravitation of the whole world—In a word, he would say, error was error,—no matter where it fell,— whether in a fraction,—or a pound,—'twas alike fatal to truth, and she was kept down at the bottom of her well as inevitably by a mistake in the dust of a butterfly's wing,—as in the disk of the sun, the moon, and all the stars of heaven put together. (145)

Although Walter has not quite articulated the "butterfly effect"—described in the popular scenario wherein the flapping of a butterfly's wings can have drastic effects on the weather—the passage seems a suggestive anticipation of it.

The interpolated "Slawkenbergius's Tale," which through sustained double entendre explains the Shandy family's obsession with big "noses," provides an apt illustration of the butterfly effect. Historians, so Slawkenbergius tells us, have ascribed grand causes to the Strasburgers' loss of their city—their refusal "to receive an imperial garrison" or the taxation that "exhausted their strength" so that they were too weak "to keep their gates shut" (271). But Slawkenbergius provides the real explanation—the Strasburgers marched out of the city "to follow the stranger's nose," thus leaving the city unprotected. As he points out, in a delightful literalization of a well-worn metaphor, "it is not the first—and I fear will not be the last fortress that has been either won—or lost by NOSES" (271). Error, so Walter Shandy would have it, "creeps in thro' the minute holes, and small crevices" (146), leading to catastrophic effects.

Sterne's text both thematizes and enacts deterministic chaos. There is, after all, a global determinism governing the text. Sterne follows a certain plan as a writer—to pen a *Life* of his hero—and whether that plan is an ad hoc one is, again, beside the point. The events that the narrator describes in the earlier volumes determine what he can say in the later ones; if he tells us at one point, for example, that the affair between Toby and Widow Wadman came to naught, he will not have them married in a later chapter. (We cannot expect the sort of trickery we find in postmodern writers, who may lead us into a Borgesian garden of forking paths by rewriting events as the narrative progresses or providing several different endings.) We even must contend with the historical determinant of Sterne's death, which apparently kept him from finishing the text. In essence, we confront an inviolable textual set-up. Yet the global determinism of the text is subject to the local randomness of social, cultural, and historical changes that result in different ways of interpreting the text—including our own era of chaos theory, which allows us to discuss it in terms of disorderly order. Our readings, as Stanley Fish reminds us, derive from the particular interpretive community out of which we operate, and chaos theory provides us with a means of accounting for the entangling of authorial intention and interpretive strategies whereby we make sense of a text.[25] With the quasi-authoritative discussion of noses, sausages, and buttonholes, Sterne points playfully to the (mis)interpretations over which he has no control.

We can also speak of the reading process as itself being determined. Although *Tristram Shandy* may be nonlinear in its approach to temporal order, we read it linearly, from first page to last. It does not invite us to plot our own reading through it, as Julio Cortázar's *Hopscotch* or Milorad Pavić's *Dictionary of the Khazars* do,—to say nothing of the virtual texts made possible by computer technology.[26] Yet local violations of the linear determination of the text can take place. Although we may sacrifice a certain amount of the sense, we can jump around, skip entire sections, return again and again to favorite passages, and wrench those passages out of context (as I do here) so that we can perform, in Barthes's terms, the "manhandling" of the text that constitutes "the work of commentary."[27] Sterne's sly directive at the outset of a new chapter that the inattentive reader "turn back" to the previous one "and read the whole chapter over again" (56)—a situation that could entrap us in a loop—and his withholding of the eighteenth and nineteenth chapters in volume nine point to his acknowledgment that even the global determinism of the linear reading process can be subverted by the local randomness of our idiosyncratic readings.

The Complexification of Time

At one point, Tristram asserts, "I almost know as little of the Chinese language, as I do of the mechanism of *Lippius*'s clockwork" (519), and this playful assertion might sum up the text itself. Sterne plays games not only with temporal ordering but also with temporal duration. As Wolfgang Iser points out, "*Tristram Shandy* must surely be the first novel to attack the substantialist concept of time."[28] Sterne's attack indeed prompted a counterattack in the form of the humorous *Clockmakers Outcry*, wherein the clockmakers throughout Britain, with their livelihoods threatened by the association that Sterne has made between clock-winding and "other little family concernments" (8), purportedly have gathered together to condemn the text.[29]

The games that Tristram Shandy plays with time are not simply a reaction to Newtonian notions, but also an anticipation and enactment of the flow of time as we have begun to understand it in the wake chaos theory. Chaos theory foregrounds the complex entanglement between external reality and internal experience in our understanding of time. It is toward this new awareness of time that Sterne's text points.

Again, Newtonian science depends on a notion of absolute time. It sets up an unbridgeable opposition between an external, absolute time and an internal time as we experience it—in Ilya Prigogine and Isabelle Stengers's terms, "the opposition, traditional since Kant, between the static time of classical physics and the existential time we experience in our lives."[30] J. T. Fraser discusses this opposition in terms of being and becoming, noting the questions it raised in the minds of eighteenth-century thinkers: "What structural part of the world is mechanical, predictable, and stable, and what part is spiritual, unpredictable, and creative? That is, how are the Eleatic categories of being and becoming divided in the nature of time?" As Fraser points out, this irreconcilable opposition has consequences: "Because of the mutually exclusive character of being and becoming, the more impressive the arguments came to be in favor of the clockwork universe, the less it was possible to find a place for man, life, and the history of life and nature."[31] In essence, there is what we think of as real time—an almost palpable entity that occurs, external to the self, measurable by the clock—and the time that we experience—subjective, idiosyncratic. As evidence of the way in which time comes to be seen as absolute according to Newtonianism, G. J. Whitrow notes that, when England shifted from the Julian to Gregorian calendar in the early part of the eighteenth century, people thought their lives were shortened, and workers rioted because they thought they had lost the wages from the days that were cut.[32] This notion of an external time, measurable by the clock, versus an internal time persists; as Barbara Adam suggests, "the time of the clock, the measure and the finite quantity, all with their implied emphasis on death, constitute central characteristics of a 'Western' cultural identity."[33] After all, more and more clocks proliferate in the world around us, appearing on our computer screens, cell phone displays, and more.

Critical assessments of *Tristram Shandy* have tended to reinforce a dualistic notion of time, making Tristram's (and *Tristram*'s) experience and exploration of time disconnected from "real" time. Dorothy Van Ghent, for example, in examining the narrator's metaleptic appeal to the reader to intervene in the story, discusses "the incongruity between the clock-time which it will take to get the two conversationalists down the stairs, and the atemporal time—the 'timeless time'—of the imagination, where the words of Toby and Mr. Shandy echo in their plenitude." She notes an essential paradox in the text: "the paradox of man's existence both in time and out of time—his existence in the time of the clock, and his existence in the apparent timelessness of consciousness."[34] Murray Krieger also deals with the temporal dualism of *Tristram Shandy*, noting that clock time functions as an enemy to both Tristram and Walter in that "the hobby-horsical world [a private reality opposed to the constraints of external reality] depends on subjective notions of time as duration rather than any linear notion of clock time."[35] Helene Moglen engages in an extended analysis of Sterne's indebtedness to Locke, pointing out that "it was Locke's view of the relation of identity and duration that provided Sterne with his fundamental theory of time."[36] Yet whereas Locke "emphasized his conviction that the rate of succession remained relatively constant in all men," Sterne, according to Moglen, emphasized "the constant clash of temporal levels" and "a gap that exists between 'private time' and 'public time' ": "Indeed, Toby and Trim's idiosyncratic war games, which always adhere closely to specific dates and historical fact, express thematically this perverse relationship of chronological and empirical time."[37] Although Calvin Thomas questions readings of *Tristram Shandy* that affirm the synchronous subjectivity of the narrator's mind over the diachronous calendar time that constitutes the objective world, he does so in order to argue that the text reveals plot as a failed plotting against time, thus reinforcing a dualistic notion of time.[38]

This is not to say that these critics have been misguided in their approaches: *Tristram Shandy* does overtly assert a duality between clock time, which Tristram associates with death, and time as it is experienced by Tristram's consciousness. However, assessing *Tristram Shandy* in light of what chaos theory has taught us about time gives us new ways of approaching this seeming duality. Through its shifts in narrative duration and intertwining of various narrative levels, the texts itself problematizes a facile opposition between real and experiential time, enacting a complexification of time that draws together the external world and internal consciousness.

Tristram Shandy was not the first eighteenth-century novel explicitly to explore the issue of narrative duration and narrative levels. In *Tom Jones*, for instance, Henry Fielding discusses the temporal compression and expansion of his material:

> When any extraordinary Scene presents itself (as we trust will often be the Case) we shall spare no Pains nor Paper to open it at large to our Reader; but if whole Years should pass without producing any thing worthy his Notice, we shall not be not be afraid of a Chasm in our History; but shall hasten on to Matters of Consequence, and Leave such Periods of Time totally unobserved.[39]

In keeping with this pronouncement, Fielding speeds up and slows down his presentation, resorting at times to brief summaries covering long periods and at others to extended scenes. He also plays with various narrative levels, blurring the distinction between author and character, author and reader, and character and reader. Yet Fielding aims to institutionalize the novel genre rather than to engage in a sustained exploration of narrativity and time, such as Sterne does. When he discusses narrative duration, Fielding wants to justify his method, not explore the connection between clock time and experiential time. When he leaps between narrative levels, he wants to stress the fictive nature of his text, not demonstrate that clock time and experiential time are inextricably entangled.

By looking closely at several episodes in *Tristram Shandy*, we can see how it enacts a complexification of time. Before doing so, however, I want to specify the narrative levels with which Sterne deals. These levels are, again, three fictional levels and three corresponding levels in the real world.[40]

The first pair of levels is the time of events. In *Tristram Shandy*, as in all written narrative, each event can be recounted at a different narrative speed, which Gérard Genette defines thus: "the relationship between a temporal dimension and a spatial dimension (so many meters per second, so many seconds per meter): the speed of a narrative will be defined by the relationship between a duration (that of the story, measured in seconds, minutes, hours, days, months, and years) and a length (that of the text, measured in lines and pages)."[41] The four narrative speeds are pause, scene, summary, and ellipsis, with the scene functioning as the only movement wherein there is an approximate equality between the time of the narration and the time of the story events. Throughout *Tristram Shandy*, Sterne pointedly calls our attention to shifts in narrative speed, often exaggerating their tendencies. As Genette makes clear, narrative speed does not correspond to some measurable duration in the real world. With regard to the scene, for example, he comments: "All that we can affirm of such a narrative (or dramatic) section is that it reports everything that was said, either really or fictively, without adding anything to it; but it does not restore the speed with which those words were pronounced or the possible dead spaces in the conversation."[42] However, although narrated events, cannot be timed as we would time events in the real world, they nevertheless point to such timeable events. Sterne indeed calls our attention to the timeability, if you will, of the events he narrates. Thus my first pairing involves the time of events as they are narrated and the time of events as they might actually take place.

My second pairing is the time of the writing. Throughout the text, narrator Tristram repeatedly draws our attention to the ongoing act of writing in which he engages, most notably in the famous statement of the Shandean time paradox discussed above. Here we have an ongoing narrating instance, a rarity in most fictional autobiographical narratives, and we have difficulty disentangling it from the time of story events, especially in that the writing of the story often becomes the story. The corresponding temporal level in reality is the actual time of writing in which Sterne engaged. Sterne anchors the text in actual dates—dates when he is presumably writing and dates of publication, as for example in the following passage at the end of volume 2: "The reader will be content for a full explanation of these matters till the next year." (154). Of course, Sterne may make up the dates, but he does actually write during a certain period, and the volumes do indeed appear in real time at set dates in the real world.

My third pairing is the time of the reading. Just as narrator Tristram repeatedly refers to the time of his writing, he also refers to the time of our reading. Indeed, he directs the continuation of the Shandean paradox to us his readers: "and consequently, the more your worships read, the more your worships will have to read" (286). The time of reading to which Tristram refers is fictive, but we read these references as we are actually engaged in the act of reading the material text *Tristram Shandy*. For Genette, our act of reading constitutes the only true measure of narrative duration:

> [C]omparing the "duration" of a narrative to that of the story it tells is a trickier operation, for the simple reason that no one can measure the duration of a narrative. What we spontaneously call such can be nothing more . . . than the time needed for reading; but it is too obvious that reading time varies according to particular circumstances, and that, unlike what happens in movies, or even in music, nothing here allows us to determine a "normal" speed of execution.[43]

As Genette maintains, although actual reading time will vary from reader to reader, the actual time of reading constitutes our consciousness of narrative duration. Through Tristram's pointed references to our reading time, Sterne emphasizes that what we regard as the duration of story events is time's passage during our reading experience.

Let us look more closely at the way in which Sterne entangles various levels of narrative duration, starting with an episode that occurs on the day of Tristram's birth—a day that it takes Sterne nearly four volumes to describe. Sterne begins with a scene in which Walter and Toby are listening to the bustle above stairs, where Mrs. Shandy prepares to give birth. After a few passages of dialogue, however, Sterne characteristically has narrator Tristram interrupt the scene with a descriptive pause—that is, a movement wherein story time (the time of events) is reduced to nothing and narrative time (the time of the telling) continues—and in the case of *Tristram Shandy*, continues and continues.[44]

This characteristic strategy, as critics have often noted, seemingly sets up an opposition between clock time, in which Toby and Walter fictively exist, and the seemingly infinite duration of human consciousness. Krieger discusses the opposition as one between the linearity of clock time (evidenced in the story events) and the circularity of Tristram's narration, and he claims that the narration is transformative: "In the hobby-horsical subjectivity of duration, however, consciousness can transform time from linear to circular."[45] In keeping with the strange-attractor model that I have invoked, I would suggest that the narration traces a spiral rather than a circle and that Sterne complicates the opposition in this sequence of the text. The narrating instance itself—the fictive time of the writing—is also subject to *a* (rather than *the*) movement of time. Narrator Tristram begins his descriptive pause with the following statement, "—Pray what was that man's name,—for I write in such a hurry, I have not time to recollect or look for it,—" (63). Soon after, commenting on the whimsicality of English characters, he notes, "that observation is my own;—and it was struck out by me this very rainy day, *March* 26, 1759, and betwixt the hours of nine and ten in the morning" (64). While the time of the story events has ground to a halt, the time of the narrating is bouncing along at a fast pace. In the childbirth episode, Toby, "whom all this while we have left knocking the ashes out of his tobacco pipe" (65), is presupposed as existing in a sort of suspended animation as Tristram "writes." Time's passage in Toby's world thus depends on the passage of time for the distracted "writer." A similar situation occurs later in the text when Tristram's mother eavesdrops upon her husband and Toby. Tristram says of her, "In this attitude I am determined to let her stand for five minutes," and we are at a loss to know whether he refers to the time she stands, the time it takes him to complete his digression, the time it takes us to read the passage, or all three (357–58). The temporal levels intertwine, undermining notions of an absolute time.[46]

Interestingly, the sense of urgency occurring in the story as Mrs. Shandy goes into labor is reflected in the narrating instance as Tristram comments upon the hurry with which he writes. Later, he enjoins a similar urgency in the time of the reading as well: "I beg, Madam, when you come here, that you read on as fast as you can, and never stop to make any inquiry about it" (76). At each local temporal level, parallel increases in speed are presupposed as occurring.

One of the most complex entanglements of the time of events, the time of narrating, and the time of reading occurs further along in the sequence when Dr. Slop, "the man midwife," arrives to assist calamitously in Tristram's birth.[47] Tristram begins by conflating the time of our reading with the time of the story events:

> It is about an hour and a half's tolerable good reading since my uncle *Toby* rung the bell, when *Obadiah* was order'd to saddle a horse, and go for Dr. *Slop* the man-midwife;—so that no one can say, with reason, that I have not allowed Obadiah time enough, poetically speaking, and considering the emergency too, both to go and come;—tho', morally and truly speaking, the man, perhaps, has scarce had time to get on his boots. (103)

Tristram explains the qualification with which he concludes his sentence in the paragraph that follows in the text: if we had timed the events in the story as if they were actually taking place in real time, Obadiah could not possibly have gone to Dr. Slop's and returned. The "hypercritic" who "is resolved after all to take a pendulum, and measure the true distance betwixt the ringing of the bell, and the rap at the door" will discover "it to be no more than two minutes, thirteen seconds, and three fifths" (103)—time herein spatialized as a measurable quantity, according to Tristram. As hypercritics, we use as our standard events as they would occur in real time, which here are seemingly speeded up impossibly in the time of the story and lengthened interminably in the time of the reading.

At this point Tristram makes his most noteworthy pronouncement about the Lockean notion of duration, setting up the opposition between it and clock time—an opposition on which so many critics have focused. Tristram would seemingly replace the clock's pendulum with another, one that follows the movement of Tristram's infinitely associative consciousness: "I would remind him [the hypercritic], that the idea of duration and its simple modes, is got merely from the train and our succession of our ideas,—and is the true scholastic pendulum,—and by which, as a scholar, I will be tried in this matter,—abjuring and detesting the jurisdiction of all the other pendulums whatever" (103). The passage is in part a bald paraphrase of Locke's statement: "the notice we take of the *Ideas* or our own Minds, appearing there one after another, is that, which gives us the *Idea* of Succession and Duration."[48] Further, Tristrams's point about "the other pendulums," implies a rejection of the fixed-point attractors common to classical physics. As proof of the validity of his notion of duration, Tristram reminds the hypercritic of all that he has narrated between the time of the bell ringing and the rap at the door:

> I would, therefore, desire him to consider that it is but poor eight miles from *Shandy-Hall* to Dr. *Slop*, the man-midwife's house;—and that whilst *Obadiah* has been going those said miles and back, I have brought my uncle *Toby* from *Namur*, quite across all *Flanders*, into *England*:—That I have had him ill upon my hands near four years;—and I have since travelled him and Corporal *Trim*, in a chariot and four, a journey of near two hundred miles down into *Yorkshire*;—all which put together, must have prepared the reader's imagination for the entrance of Dr. *Slop* upon the stage. (103-4)

The duration of the narration, which has included a long flashback dealing with Toby's past history, is translated into the reading time that prepares the reader for Slop's appearance, which would otherwise indeed be improbably untimely. The duration of the narrating and reading thus supersedes the duration of events in real time.[49]

In a playful turn of the screw, however, Sterne makes clear that real time still pertains—that no temporal laws of probability were broken after all. Tristram explains to the hypercritic: "I then put an end to the whole objection and controversy about it all at once,—by acquainting him, that *Obadiah* had not got above three-score yards from the stable-yard before he met with Dr. *Slop*" (104). All temporal levels, including that of real time, are valid in the world of *Tristram Shandy*.

The conflation of temporal levels becomes particularly exacerbated in volume 7. Here the temporal progression of the narrating instance—the apparent time of the writing—replaces the temporal progression of the story of Tristram's early life, with Tristram describing his "present-day" travels across Europe as he flees death: "There's FONTAINEBLEU, and SENS, and JOIGNY. . . . I might as well talk to you of so many market-towns in the moon, as tell you one word about them: it will be this chapter at the least, if not both this and the next entirely lost, do what I will—" (510). Tristram seemingly writes to the moment during his travels.

A few pages later, however, Tristram makes the following comment, "Now this is the most puzzled skein of all," a comment that aptly sums up the entanglement of temporal levels and narrative threads that will follow (515). He explains the puzzle thus:

> I have brought myself into such a situation, as no traveller ever stood before me; for I am at this moment walking across the market-place of *Auxerre* with my father and my uncle *Toby*, in our way back to dinner—and I am this moment also entering *Lyons* with my post-chaise broken into a thousand pieces—and I am moreover this moment in a handsome pavilion built by *Pringello*, upon the banks of the *Garonne*, which Mons. *Sligniac* has lent me, and where I now sit rhapsodizing all these affairs. (515–16)

The initial narrating instance, occurring as Tristram travels, has become story entirely, replaced by a new narrating instance wherein Tristram reflects back on those travels. Presumably, his sojourn on the banks of the Garonne is subject to temporal progression as well. Genette refers to the uniqueness of Sterne's proceeding: "the fictive narrating of that narrative, as with almost all the novels in the world except *Tristram Shandy*, is considered to have no duration.... One of the fictions of literary narrating, perhaps the most powerful one, because it passes unnoticed, so to speak—is that the narrating involves an instantaneous action, without a temporal dimension."[50] By violating this implicit code of narrating, Sterne compels us to contemplate the fact that we may regard time as advancing at different rates at different levels simultaneously, thereby problematizing the priority of one real time against which all others stand as imitations.[51]

Certainly Sterne's games with time are audacious. But in what way might they be considered as enacting chaotic time? And, perhaps more to the point, of what relevance are such games to our own understanding of time and narrative? After all, *Tristram Shandy* is just a fiction, and fiction writers are at liberty to do anything they want with time. Why should we think that a fictional narrative can teach us anything about what time really is or that a new understanding of time can enhance our understanding of narrative structure? As Bastiaan van Fraassen remarks, however, an "intimate relation between narrative time and physical time" exists: "[T]he constitution of time in our construction of the real world is not different in essential character from the constitution of time by the reader in his construction of the narrated world as he reads the text."[52] This reciprocity between time and narrative entails that, when our concept of time changes, so does our concept of narrative—and vice versa.

Whereas Newtonian science presupposes an absolute time, chaos theory makes us aware that different local levels, or scales, of time may be operating at different rates simultaneously. In a provocative discussion between humanists and physicists on the International Society for the Study of Time listserv, Paul Harris lists several possible properties of "chaotic time," including the following: "We look for patterns across different scales or levels rather than through the years (or along a time line). There is no global law or single external time parameter or measure to use to tick time off . . . each local event has its own law."[53] In a response, Thomas Weissert elaborates upon the notion of temporal levels: "Time means different things at different levels, and if we throw out the box idea and focus on the unfolding of time, the process, then it is operating at all levels together and differently, yet similarly."[54] Chaos theory makes us aware of "the creative emergence of increasingly complex temporalities," in Fraser's terms.[55] *Tristram Shandy* performs these complex temporalities.

John Casti speaks of such a complexification of time as "the manifestation of relations between events." Contrasting this conception with the Newtonian notion of time, he argues: "[W]e have to reject the idea that time is like some continually flowing stream of moments waiting to be filled by events. . . . What's needed instead is the version espoused by Aristotle . . . where we understand change to be the experience of some structure of events."[56] Claiming that assessing events along a Newtonian linear time line can "warp the natural geometry of events," Casti takes what he calls "a multidimensional view of time," using the event "house" as an illustration. In doing so, he demonstrates how the same period of time may be observed differently at different levels: "[T]he level-N observer sees a 7 event occur in 28 days; for the level-N + 1 observer, the same event takes 1792 days, or approximately 5 years."[57] *Tristram Shandy* both explores how time may be regarded as functioning differently at different levels, none of which has absolute priority, and provides an interactive demonstration of the phenomenon.

Finally, I want to look at a highly speculative but, in light of the above, credible proposition about chaotic time, one that shares clear affinities with the assumptions about time made in *Tristram Shandy*. Addressing his often audacious interdisciplinarity and his tendency to find contemporary relevance in out-of-date authors and texts, Michel Serres proposes a theory of chaotic time: "Time does not always flow according to a line . . . nor according to a plan but, rather, according to an extraordinarily complex mixture, as though it reflected stopping points, ruptures, deep wells, chimneys of thunderous acceleration, rending, gaps—all sown at random, at least in a visible disorder."[58] Tristram provides us with illustrations of the narrative trajectories of the first five volumes of his work. Opposed to the straight line he promises—in vain—to follow, they appear to be riddled with "stopping points, ruptures, deep wells," and so forth. According to Serres, "Time doesn't flow; it percolates. . . . time flows like the Seine, if one observes it well. All the water that passes beneath the Mirabeau Bridge will not necessarily flow out into the English Channel; many little trickles turn back toward Charenton or upstream."[59] Indeed, the strange temporality of *Tristram Shandy* might be likened to a flow that has become turbulent, that has begun to percolate, Sterne thereby demonstrating his incipient awareness that there is more to time than the Newtonians would have us believe. Sterne's narrative may be fictional, but it functions to illuminate the real, enabling us to understand the complex nature of perceived time to which chaos theory draws our attention.

Through its exacerbation of disorderly order, *Tristram Shandy* serves as a paradigm text for discussions of narrative structuration and chaos theory. In Viktor Shklovsky's famous formulation, it may indeed be "the most typical novel of world literature."[60] Through its very strangeness, it draws our attention to the chaotic element that may be inherent in narrative itself. Within the bounded state-space of his text, Sterne playfully enacts complex temporalities and puts in motion a trajectory that promises to evolve infinitely as it bounces between the unstable attracting points of sex and death.

Despite his attempt to "mend" himself, Sterne does not give us the straight line of the linear text. He understands only too well that it is indeed "the line of GRAVITATION" (474–75) that leads to the blank, black page, which comes at the end of every *Life*.

CHAPTER 3

Narrating the Workings of Memory: Iteration and Attraction in *In Search of Lost Time*

> *If we study the history of science we see produced two phenomena which are, so to speak, each the inverse of the other. Sometimes it is simplicity which is hidden under what is apparently complex; sometimes, on the contrary, it is simplicity which is apparent, and which conceals extremely complex realities.*
>
> —Henri Poincaré, *Science and Hypothesis*

> *If the object of analysis is indeed to illuminate the conditions of existence—of production—of the text, it is not done, as people often say, by reducing the complex to the simple, but on the contrary by revealing the hidden complexities that are the secret of the simplicity.*
>
> —Gérard Genette, *Narrative Discourse*

> *We now see a hierarchical structure of great complexity emerging gradually.*
>
> —Ivar Ekeland, *Mathematics and the Unexpected*

In *Time and the Novel*, A. A. Mendilow notes that Laurence Sterne "clearly anticipates the time-shift technique of the twentieth century, in his search for a truer notation of the processes themselves of experience."[1] More specifically, *Tristram Shandy* anticipates Marcel Proust's *In Search of Lost Time* (*Recherche du temps perdu*). Jean-Jacques Mayoux explicitly makes the connection, saying of Sterne that "his perhaps crowning inspiration has been to fling himself into his own stew in the guise of the narrator, Tristram, the like of whom had never been seen, and was not be seen again until Proust's Marcel."[2] In each text, the writer subverts autobiographical conventions in an attempt to retard time's forward march and to recapture lost time.

Whereas Sterne crafts a fictional autobiography, Proust gives us instead what we might call a fictionalized autobiography. Indeed, at one point, the narrator states that "[T]here is not a single incident which is not fictitious, not a single character who is a real person in disguise, in which everything has been invented by me in accordance with the requirements of my theme"[3] The overall deterministic movement of autobiography from birth to death (specifically, the death of the narrated self into the narrating self) informs both Sterne's and Proust's texts. Both writers suggest, however, that they could expand the possibilities between those two points indefinitely. They thus demonstrate that the deterministic linear structure common to autobiography reflects neither the workings of the world nor of the mind.

Granted, compared with a text such as *Tristram Shandy*, *Search* may seem downright traditional. Although it takes a few pages to get there, the narrative has what we might call a real beginning—the inauguration of the narrator's quest to become an artist—and it reaches a definite conclusion—the narrator's assumption of his artistic vocation.[4] Yet, as Gérard Genette's extensive analysis of *Search* in *Narrative Discourse* reminds us, we are dealing with a narrative that consistently violates the traditional codes that it would ostensibly uphold, and the seeming simplicity of its autobiographical structure "conceals extremely complex realities," to borrow Henri Poincarés's phrase.[5] Genette's structuralist discussion of Proust's technique is exemplary, but we need to move beyond it to appreciate the text's narrative dynamics. Insofar as it is possible, Proust aims to create a work that is seemingly, to borrow one of Proust's own phrases, "perpetually in the process of becoming" (ML 6: 522/G 2396)—a work that within the static space of its pages will convey temporal flux.[6]

Throughout *Search*, Proust conveys an incipient awareness of key features of deterministic chaos. Moreover, Proust's violations of the traditional codes of frequency and order give the text's structure a quality of bounded randomness similar to what we find in a chaotic system.[7] To an unprecedented degree, Proust draws on what Genette terms the "iterative mode" of frequency—that is, taking several similar events and synthesizing them so as to recount them only once in the text.[8] This iterative mode creates a global pattern from the various local random fluctuations that occur, an imaginative projection of potentially infinite occurrences. The emergent structure, analogous to that generated when nonperiodic trajectories fall onto a strange attractor, enables Proust to show how the random events of human lives achieve meaning and reality through the synthesizing power of memory. An attractor-structure also represents the text's overall narrative trajectory, which oscillates between the narrator's desire for amorous consummation and his desire for an artistic vocation. This oscillation occurs at both local and global levels, exemplifying the property of similarity across scale found in chaotic systems. Although the overarching narrative trajectory progresses linearly, it consistently turns back upon itself to fill in more information, leaving us with a sense that the text could expand indefinitely as the narrator searches for a fulfillment he can never achieve. Ultimately, however, the narrator's epiphany at the Guermantes matinée enables him to vary the parameters of this system, and the strange attractor collapses into a classical fixed-point attractor, signifying the narrator's readiness to assume his artistic vocation by recounting his life. As Proust demonstrates, it is by figuring forth a time that never was and by playing against time's fulfillment that one may come to know "real life."

Proust, Poincaré, and "Unforeseen Perturbations"

With its *ab ovo* beginning, inconclusive conclusion, and "digressive/progressive" narrative trajectory, *Tristram Shandy* questioned the deterministic thinking of classical science and fought against its own "death" in the process. Whereas Sterne wrote at a time when the deterministic view of the natural world instigated by Newtonian science was assuming ascendancy, Proust launched his writing career in the late nineteenth century, when classical determinism had reached its apogee, as Ivar Ekeland explains: "By the end of the nineteenth century, it was well entrenched, not only within the scientific community but also among philosophers and the public at large. It held that the laws of nature were known, or would be known, and that the world was deterministic, so that predicting the future from the present was just a matter of computation."[9] As Ekeland also makes clear, such thinking was about to be undermined by the French mathematician Henri Poincaré, the originator of dynamical systems theory, who would "initiate the critical analysis of classical determinism, thereby opening the modern era."[10]

It is tempting to link Marcel Proust, one of the first literary modernists, with Poincaré, the inaugurator of the modern era of science. Certainly, Proust knew Poincaré's work, and many passages in *Search* point toward his understanding of key features of deterministic chaos. In *The Guermantes Way* (*Le Côté de Guermantes*), when the young aristocrat Robert de Saint-Loup theorizes about "the richness of the world of possibilities as compared with the real world" (ML 3: 148/G 834), he refers directly to Poincaré: "To go back to our philosophy book; it's like the rules of logic or scientific laws, reality conforms to them more of less, but remember the great mathematician Poincaré: he's by no means certain that mathematics is a rigorously exact science" (ML 3: 149/G 834). One wonders if, when writing these words, Proust was thinking of a passage such as the following, wherein Poincaré calls into question the laws of classical physics:

> Have we any right, for instance, to enunciate Newton's law? No doubt numerous observations are in agreement with it, but is not that a simple fact of chance? and how do we know, besides, that this law which has been true for so many generations will not be untrue in the next? To this objection the only answer you can give is: It is very improbable. But grant the law. By means of it I can calculate the position of Jupiter in a year from now. Yet have I any right to say this? Who can tell if a gigantic mass of enormous velocity is not going to pass near the solar system and produce unforeseen perturbations?[11]

Poincaré here suggests that an unexpected event can undermine the absolute predictability of physical phenomena. A passage such as this one, which paves the way toward our contemporary awareness of deterministic chaos, has affinities with the narrator's discussion of the unpredictability of the future: "But we picture the future as a reflexion of the present projected into an empty space, whereas it is the result, often almost immediate, of causes which for the most part escape our notice" (ML 5: 430/G 1844). Those causes that "escape our notice" are analogous to Poincaré's "unforeseen perturbations"—that is, the sensitive dependence on initial conditions whereby the microscopic affects the macroscopic.

Although we cannot know whether Proust consciously drew on Poincaré's theories as he formulated his great masterpiece, we do know that he was aware of contemporary scientific thought. The narrator, in fact, makes a connection between the work of the artist and that of the scientist: "The impression is for the writer what experiment is for the scientist, with the difference that in the scientist the work of the intelligence precedes the experiment and in the writer it comes after the impression" (ML 6: 276/G 2273). The writer thus proceeds scientifically, examining the materials—the impressions—at his or her disposal in the same way that a scientist would observe an experiment. Indeed, the work of art is, like Alexander Pope's naturalized aesthetic criteria, "discovered, not devised": "I had arrived then at the conclusion that in fashioning a work of art we are by no means free, that we do not choose how we shall make it but that it pre-exists us and therefore we are obliged, since it is both necessary and hidden, to do what we should have to do if it were a law of nature—to discover it" (ML 6: 277/G 273). Like the scientist, the writer brings to light some facet of the natural world. The writer discovers a preexistent pattern that dictates how the book will be shaped.

At one point, the narrator draws on mathematical language, pondering "what logarithmic table" (ML 5: 485/G 1874) describes Albertine's body—although elsewhere he notes, "no mathematical process would have enabled one to convert Madame d'Arpajon and Madame de Montpensier into commensurable quantities" (3: 782/G 1183). Even so, the narrator clearly regards his characters in terms of mathematical properties. Assuming that the narrator's aesthetic pronouncements dovetail with Proust's own beliefs, we can argue that Proust regarded artistic and scientific endeavor as analogous, with Proust's aesthetic notions arising from the same ferment of ideas from which Poincaré's theories came.

Throughout the text, Proust's descriptions suggest an awareness of the features that we have come to associate with deterministic chaos.[12] At one point, the narrator discusses how a minor incident changed two destinies: "In this way, accidentally and absurdly, a minor incident (in this case the juxtaposition of Albertine and Saint-Loup) has only to be interposed between two destinies whose lines have been converging towards one another, for them to deviate, stretch further and further apart, and never converge again" (ML 4: 684/G 1583). The description recalls that of the orbits of a chaotic attractor, which, despite having similar initial conditions, "diverge exponentially fast and so stay close together for only a short time."[13] Proust once again demonstrates an incipient awareness of sensitive dependence on initial conditions, which can cause two trajectories either to converge or to diverge. Deterministic chaos rules interactions between people such as Albertine and Saint-Loup.

Furthermore, when the narrator describes his own interactions with the various characters in *Search*, the description evokes the structure of a strange attractor as it emerges in state space:

> How often had all these people reappeared before me in the course of their lives, the diverse circumstances of which seemed to present the same individuals always, but in forms and for purposes that were shifting and varied; and the diversity of the points in my life through which had passed the thread of the life of each of these characters had finished by mixing together those that seemed the furthest apart, as if life possessed only a limited number of threads for the execution of the most different patterns. (ML 6: 415/G 2543-44)

In effect, the narrator describes "the baker's transformation," in which the stretching and folding that takes place enables exponentially diverging orbits to be encompassed within fixed bounds. Although the narrator repeatedly encounters "the same individuals," at times the destinies of certain individuals appear far apart or "stretched" and at others close together or "folded," a situation in keeping with the bounded randomness of Proust's text.

The narrator's later elaboration of this theme indeed suggests the bounded randomness of the strange attractor, whose trajectory crosses and recrosses itself, increasingly filling up state space:

> But the truth, even more, is that life is perpetually weaving fresh threads which link one individual and one event to another, and that these threads are crossed and recrossed, doubled and redoubled to thicken the web, so that between any slight point of our past and all the others a rich network of memories gives us an almost infinite variety of communicating paths to choose from. (ML 6: 504/G 2388)[14]

During the time when the narrator is in love with Gilberte, she speaks to him of Albertine, whom he will come to love; Albertine speaks of her close connection with the friend of Mlle Vinteuil, whom the narrator had long before observed engaging in a profane ritual; Vinteuil's Sonata serves as a leitmotif for Swann's love affair with Odette, the mother of Gilberte. Like the trajectory of the strange attractor, traversing state space to link distant points, the life trajectory of the narrator creates unexpected links.

The narrator makes this point especially explicit when he speaks of Mlle de Saint-Loup. On this child of Gilberte Swann and Robert de Saint-Loup, the narrator muses, all the trajectories of his life converge: "Numerous for me were the roads that led to Mlle de Saint-Loup and which radiated around her" (ML 6: 502/G 2387). After enumerating these connections, he concludes: "And to complete the process by which all my various pasts were fused into a single mass Mme Verdurin, like Gilberte, had married a Guermantes" (ML 6: 505/G 2388). As the narrator's words suggest, Mlle de Saint-Loup is the attractor par excellence, drawing seemingly disparate points toward her. Most significantly, she draws together the Méséglise and Guermantes Ways—those actual and symbolic destinations that enable the emergence of the text's strange-attractor structure.

Iteration, the Iterative, and the Iterated Walks

In linking a mathematical procedure with Proust's affinity for a particular grammatical tense (the *imparfait* or imperfect), I am not simply exploiting an etymological connection. Obviously, Proust is not feeding sentence formulas into a computer to come up with a solution—although one can envision an interesting speculative fiction à la Borges or Calvino along these lines. However, the dynamicist's iterations and Proust's use of the iterative mode act analogously: each involves drawing on numerous variables to come up with an overall system trajectory, which approximates a true solution. Indeed, for both the dynamicist and Proust, this method is the only access to the true solution.

In mathematics, iteration refers to performing a calculation repeatedly. For a dynamicist, such iteration entails calculating the state of a dynamical system as it evolves through time. The process is recursive—that is, the output of a particular calculation becomes the input for the next. Superfast computers have enabled dynamicists to perform multiple iterations and thus map deterministic chaos. In fact, so many iterations are involved in this mapping that it would be impossible for one to perform them without computers.[15]

Robert Shaw's landmark study of the dripping faucet illustrates this process (see chapter 1). In order to map the system's transition to chaos when the flow rate was increased, Shaw needed to perform numerous iterations of three variables: the position, mass, and velocity of the drops. James Gleick describes Shaw's iterative procedure: "As each drop fell, it interrupted a light beam, and a microcomputer in the next room recorded the time. Meanwhile Shaw had his three arbitrary equations [three differential equations modeling the drops' interaction] up and running on the analog computer, producing a stream of imaginary data."[16] As the computer performed its iterations, this "imaginary data" took shape as a chaotic or strange attractor—in this case, a funnel attractor. An iterative process such as Shaw performed produces an evolving pattern in state space, a global pattern emerging from local randomness.

Significantly, the process of iteration leads to only an approximation of a true solution. When mapping a dynamical system, a dynamicist must choose at which time intervals to take measurements, for not all points can be represented.[17] Decreasing the time interval leads to greater accuracy, but a chaotic system poses challenges, as Thomas Weissert explains: "In regions of 'deterministic chaos,' in which the trajectory changes its course rapidly, the numerical procedure generates noise in the simulation regardless of the time-step size. This level of noise grows until the signal—the true solution—gets lost in it."[18]

Furthermore, because mapping requires that we "choose some finite numerical grid of precision," our accuracy is limited.[19] In effect, the choice we make affects the results obtained, and we can never actually reach a true solution. It is no wonder that, in *The Art of Modeling Dynamics*, Foster Morrison instructs the would-be dynamicist, "Always remember that a model is not reality, but something that imitates reality at a certain scale."[20] Ultimately, as Weissert suggests, we face "the randomness associated with trying to connect a physical system to a numerical initial condition."[21] We cannot obtain the thing itself.

Although the model generated by the process of iteration may be only an imitation of reality, an approximation of a true solution, it nevertheless confers an identity upon the system. Out of the local randomness a global pattern evolves, which gives us our only access to the true solution.[22] Significantly, this identity is an emergent one. The strange attractor, once set going on the computer screen, will continue to evolve, its trajectory moving through a potentially infinite number of different states.

Similarly, Proust's use of the iterative mode of frequency enables him to create a global pattern for the past from the various local events that occur, a global pattern that is not fixed but evolving. Thereby does the past remain alive. What is at issue here is not whether Proust himself captured his own lost time. As we know, there was no Combray: Proust merged his memories of childhood days spent in Illiers and Auteuil in order to create this fictional city.[23] The madeleine itself—the most famous cookie in literature—was initially just a dry biscuit.[24] When we examine the evolution of the text itself (its inception in *Jean Santeuil* and the *Contre Saint-Beuve* essay, its various permutations in the notebooks), we can see that Proust was concerned not with evoking his own past (although certainly that past resonates in the text), but with demonstrating, through a careful crafting of episodes and a deliberate attention to style, the mapping and modeling process whereby a past might be recaptured. Thus the plot of *Search* recounts the narrator's efforts to achieve the structure and style that would enable him to recapture the past, presumably successful efforts if we take "his" book as evidence.

The iterative sections of *Search* are set in an indefinite—even an unfinished—past shaped by Proust's use of the imperfect tense. Proust describes, as Genette points out, "in the French imperfect tense for repeated action, not what *happened* but what *used to happen* at Combray, regularly, ritually, every day, or every Sunday, or every Saturday, etc."[25] Rather than recounting what happens on a particular day during the Easter holidays, the narrator prefers to recount what happens over the course of many days, thus to some extent giving us a sense not only of what used to happen but of what could happen—of the endless possibilities precluded by narrating one time what happens once, the so-called singulative mode of frequency.[26] The overall rhythm of the text consists of an oscillation between the iterative and the singulative, the ritualistic and the exceptional. Particularly noteworthy instances of the iterative occur in the "Combray" section of *Swann's Way* (*Du côté du chez Swann*) when the narrator describes the two walks that he used to take with his parents during his childhood visits to Combray and in *Sodom and Gomorrah* (*Sodom et Gormorrhe*), when he describes the railway stops between Balbec and La Raspelière.

Each of the two walks—the brief one along the Méséglise Way (Swann's Way) and the longer one along the Guermantes Way—comprises a trajectory that winds through Combray and its environs over a certain time period. It is tempting to make a connection between these trajectories and that generated by the strange attractor. After all, the descriptions of these walks spatialize time or temporalize space, to some extent. But the state space in which a strange attractor evolves is not the same as the actual space of the French countryside, and a walk is only a dynamical system in the most unscientific of senses. Nevertheless, Proust's use of the iterative mode to describe the walks confers upon them a bounded randomness similar to that of the strange attractor: they are globally determined but subject to local randomness. The Méséglise Walk as Proust depicts it no more represents a "real" walk (within the contract of the fiction) than the strange attractor represents a real object. Instead, like the strange attractor, it is an imaginative projection formed by the trajectories of a potentially uncountable number of walks. For Proust, this *figurative* walk has greater potential for enabling the regaining of lost time than the most concrete, particularized of walks could do.[27]

Over the course of a decade or so during his youth, the narrator accompanies his parents along the two paths. Clearly, although the family members may confine their walks to two fixed routes, during each walk, random events occur. One day they might return late to find the servant Françoise anxiously on the lookout, another day a crow might be borne into the air on a gust of wind, and yet another day the narrator might yearn for "a peasant girl to embrace" (ML 1: 221/G 131). In some cases, the events are noteworthy enough to warrant a singulative scene, indicated in the French text by the *passé composé*: for example, when the narrator encounters the scornful Gilberte at Tansonville or when he spies on Mlle Vinteuil and her lesbian friend as they profane M. Vinteuil's picture. Sometimes, the events are pseudo-iterative, a seemingly singulative event presented in the imperfect tense.[28] For the most part, however, local variations are absorbed into the iterated walks. A feedback loop, such as the one that occurs during the process of mathematical iteration, thus contributes to the overall pattern. Significantly, neither of the iterated walks does, in fact, really occur at all; the walks are manifestations of an emergent structure containing some—but not all—of the local variations that take place during each.[29]

Another significant use of iteration occurs in the last third of *Sodom and Gomorrah*. Recalling his journeys to La Raspelière during the autumn, the narrator imaginatively revisits the train stations along the route of the transatlantic train (the T.S.N.) to spur his memory. The present of his recollections merges with the past of his experiences: "But already my memories of what

I was told about all this are giving place to others, for the T.S.N., resuming its slow crawl, continues to set down or take up passengers at the succeeding stations" (ML 4: 656–67/G 1569). An interestingly *Shandy*-like conflation of narrative levels occurs both here and in an earlier passage in which the narrator excuses himself for not elaborating on certain incidents concerning the violinist Morel: "There were others, but I confine myself at present, as the little train halts and the porter calls out 'Donciéres,' 'Grattevast,' 'Maineville' etc., to noting down the particular memory that the watering-place or garrison town recalls to me" (ML 4: 648/G 1564). In each passage, the present of the narrating instance is conflated with the past of the train's journey, as if the narrator were not only recalling his journeys on the T.S.N. but also participating in them again, which, if we accept the underlying premise about the recovery of lost time, he would, in effect, be doing.

During the T.S.N. episode of the text, the iterative mode enables the recovery of memory. Unlike in the "Combray" passages, which deal with walks during different seasons and at different times of the day, in this episode, the one T.S.N. journey undertaken by the narrator in his mind stands for all the journeys to La Raspelière. Each of the various stops along the route triggers a particular memory, such as Charlus's fabrication of a duel to compel Morel to return to him or the narrator's refusal to greet Bloch's father because he fears leaving Saint-Loup and Albertine together. Like the "Combray" instances, however, the T.S.N. episode absorbs the singulative into the iterative, providing us with a synthetic (in all senses of the word) journey.

For Proust, the emergent structure of the iterated walks or train journeys more aptly captures lost time than the individualized walk or train journey would, as his narrator's aesthetic pronouncements demonstrate. To a large extent, *Search* serves as a rumination upon the processes that brought it into being. Whether he is musing on the gestures of Berma, on Bergotte's exhausting style, or on the effect of the Vinteuil Sonata on its listeners, the narrator compels us to regard his ruminations in light of the text's underlying aesthetic. Whether he is pondering the workings of memory, the inexorability of time, or the nature of reality, he compels us to consider all these discussions in light of the text's underlying philosophy. In essence, the text provides the code for its own reading.

As the narrator explains, a fixed image is a dead image. After hearing Bergotte praise Berma's arm gesture in *Phèdre*, he laments that he cannot synthesize this impression with his own disenchanted impression: "But all that I retained of Berma in that scene was a memory which was no longer susceptible of modification; as meager as an image devoid of those deep layers of the present in which one can delve and genuinely discover something new, an image on which one cannot retrospectively impose an interpretation that is not subject to verification and objective sanction" (ML 2: 184/G 446). Fixing a past experience in memory takes it out of context and robs it of "those deep layers of the present"—those possibilities lodged in the "state space" of the mind. Hence the narrator's reluctance to read the letter he receives from Mme de Stermaria: "I hesitated for a moment before looking to see what Mme de Stermaria had written, which as long as she held the pen in her hand might have been different, but was now, detached from her, an engine of fate pursuing its course alone" (ML 3: 536/G 1046). The narrator hesitates to read Mme de Stermaria's letter because it will determine the fate of his relationship with her. Determination is, in fact, fatal, precluding new, alternative developments—just as occurs with a classical deterministic system, such as the pendulum, whose trajectory can lead to only one conclusion.

We see a similar theme emerge in the narrator's discussion of his projected trip to Italy at the time of his infatuation with Gilberte. He explains that, by fixing dates for his Italian sojourn, his father confers a reality upon the Italian cities that they had so far lacked:

> They became even more real to me when my father, by saying, "Well you can stay in Venice from the 20th to the 29th and

reach Florence on Easter morning," made them both emerge, no longer from the abstraction of Space, but from that imaginary Time in which we place not one journey at a time but others simultaneously, without too much agitation since they are only possibilities—that Time which reconstructs itself so effectively that one can spend it again in one town after one has already spent it in another—and assigned to them some of those actual calendar days which are the certificates of authenticity of the objects on which they are spent, for these unique days are consumed by being used, they do not return, one cannot live them again here when one has lived them there. (ML 1: 558/G 316)

The narrator here makes a distinction between the time of fixed dates and the imaginary time of the indefinite, the consumed day and the day that one can live again and again. The consumed day's course has already been determined, and that time is irrecoverable—the pendulum has come to rest on a fixed point.[30]

Lost time can be found again only in the realm of possibility. Of course, the phrase "imaginary time" suggests that the lost time for which the narrator searches is purely illusory. Yet the imaginary time of unfixed days partakes of a reality that precise dates, particular days, cannot:

> [B]ecause I had refused to savour with my senses this particular morning, I enjoyed in imagination all the similar mornings, past or possible, or more precisely a certain type of morning of which all those of the same kind were but the intermittent apparition which I had at once recognized.... This ideal morning filled my mind full of a permanent reality, identical with all similar mornings, and infected me with a joyousness which my physical debility did not diminish. (ML 5: 24/G 1621–22)[31]

Through the descriptions of the iterated walks and the railway journey, Proust attains this ideal, an ideal designed to evoke a living past that is not so much "past or possible" as past *and* possible. Recall that the trajectory orbiting around a strange attractor signifies the past of the dynamical system as well as points toward its future—a future rife with possibilities in the realm of state space. Consider also Poincaré's comments regarding the mathematician: "to prove even the smallest theorem he must use reasoning by recurrence, for that is the only instrument which enables us to pass from the finite to the infinite."[32] Proust's use of the iterative is his own version of "reasoning by recurrence," enabling him to pass from the finite situation to infinite possibilities that for him compose the truth of experience.

For Proust, the ideal is the real. It partakes of "permanent reality." As the narrator comments, "[R]eality takes shape in the memory alone" (ML 1: 260/G 151). Memory synthesizes in order to convey reality: "And more even than the painter, the writer, in order to achieve volume and substance, in order to attain to generality and, so far as literature can, to reality, needs to have seen many churches in order to paint one church and for the portrayal of a single sentiment requires many individuals" (ML 6: 316–17/G 2294). Just so does the dynamicist engage in an iterative process to produce the emergent pattern of a chaotic system, thus presenting a figure that enables us to glimpse the system's dynamics. Out of the iterations of many childhood walks, of many journeys between Balbec and La Raspilière, an emergent structure takes shape, and the synthesizing memory can thereby evoke the reality of a time lost.

Méséglise, Guermantes, and Similarity across Scale

The narrator's pronouncement that the writer must see "many churches" is key to our understanding of the structuration of *Search*. The text's narrative trajectory, like a butterfly attractor, oscillates between two attracting points, continuously revisiting past sites and filling in new information. It is thus that the writer can "achieve volume and substance." Throughout the text the narrator's desire for amorous consummation and desire for artistic vocation remain unfulfilled, but this deferral enables the narrator to arrive at the truths that will eventually give him the materials for his novel and consequently lead to the collapse of the strange attractor onto a fixed-point attractor.

Whereas Sterne consistently and explicitly subverts the birth-to-death linear chronology of autobiographical narrative, Proust's method is more subtle. Despite the many chronological shifts in *Swann's Way*, including the extended flashback of "Swann in Love" ("Un amour de Swann"), once we reach "Place-Names: The Name" ("Nom de pays: Le nom"), the overall narrative trajectory of *Search* proceeds linearly through the final line of *Time Regained* (*Le temps retrouvé*), moving from the narrator's adolescence to his renewed and presumably permanent commitment to his artistic vocation, articulated during the Guermantes reception.[33] Given that Proust composed the conclusion of *Search* at the same time as its beginning, we must assume that he planned that what happened in the beginning would lead inevitably to the ending with an almost Aristotelian plot logic.[34]

As has often been recounted, when *Swann's Way* was published in 1913, Proust conceived it as part of a trilogy, which was already substantially complete. Jean-Yves Tadié points out, however, that "[Proust's] experiences between June 1913 and the summer of 1914, followed by the suspension of all publishing at Grasset's on account of the war, would alter all existing plans and, in a totally unexpected way, make the work double in size—it would expand from 1,500 to 3,000 pages in eight years."[35] Marion Schmid notes, "This enormous expansion happened essentially at the centre of the novel, whilst the beginning (*Du côté de chez Swann*) and the ending (*Le temps retrouvé*) largely preserve their original outline."[36] Because of the contingent circumstance of the war, Proust continued over an eight-year period to revisit, revise, and, most importantly, enlarge what he had already written. Yet this process seems intrinsically connected to his overall plan for *Search*. Within the constraints of a fixed beginning and ending, the attractor structure of the text emerges. The "dynamical system" of Proust's text is both deterministic with regard to the main plan and chaotic with regard to the way that the plan will be realized.[37]

Like many chaotic systems, *Search* manifests similarity across scale. "Combray I" enacts in miniature the key action of the entire work—the narrator's fall into disillusionment and artistic sterility, followed by a revivifying instance of involuntary memory. At the chapter's conclusion, the image of the formless paper bits, submerged in water and then blossoming into a miniature village, symbolizes his desire to recreate Combray, a desire that will be reiterated at the conclusion of the entire text.[38] The story of Swann's ill-fated love for Odette, detailed in the self-contained "Swann in Love," is reiterated on a grander scale in the story of the narrator's ill-fated love for Albertine. Throughout *Search*, what happens on a local level occurs on a global one as well, this similarity across scale reinforcing the deterministic chaos of the text.

The two walks serve as an especially significant instance of similarity across scale. They represent unattainable attracting points at both local and global levels. Although the narrator's family makes its predetermined circuit along both the Méséglise and the Guermantes Ways, they never attain the ultimate goal of each walk:

> As for Guermantes, I was to know it well enough one day, but that day had still to come; and, during the whole of my boyhood, if Méséglise was to me something as inaccessible as the horizon, which remained hidden from sight however far one went, by the fold of a landscape which no longer bore the least resemblance to the country round Combray, Guermantes, on the other hand, meant no more than the ultimate goal, ideal rather than real, of the "Guermantes way," a sort of abstract geographical term like the North Pole or the Equator or the Orient. (ML 1: 188–89/G 113)

The ultimate destinations Méséglise and Guermantes are the unattainable attracting points around which the trajectories of the iterated walks orbit—attracting points at the local level. During the narrator's boyhood, the Méséglise Way is the site of erotic desire in all its permutations—where he falls hopelessly in love with the scornful Gilberte Swann and where he witnesses the profane rites of Mlle Vinteuil and her friend. The Guermantes Way is the site of artistic aspiration—where he envisions walking with Mme. de Guermantes and telling her about the poems he intends to write and where he composes the little fragment on the Martinville steeples.

Functioning literally at the local level, the two destinations also function symbolically at the global level. At the crucial Guermantes reception, the narrator comes to the following realization:

> And indeed my whole social life, both in the drawing-rooms of the Swanns and the Guermantes in Paris and also that very different life which I had led with the Verdurins in the country, was in some sense a prolongation of the two ways of Combray, a prolongation which brought into line with one way or the other places as far apart as the Champs-Élysées and the beautiful terrace of La Raspilière. (ML 6: 503/G 2387)

The narrator thus makes explicit the figurative import of the literal walks. The state of unrequited love and the observation of amorous perversions that the young narrator experiences along the Méséglise Way will become defining features of his adulthood. The aesthetic conversations and artistic productions of which he dreams when he wanders the Guermantes Way as a boy will also define the man that he will become. The two Ways serve as the unattainable attracting points around which the overall plot trajectory orbits. At the global level, these points attract and repel the narrative trajectory, creating the text's strange-attractor structure. Throughout *Search*, the narrator oscillates between erotic desire and artistic aspiration, each, like the ultimate destination of the two ways, seemingly out of his reach.

Just as the narrator's family would alternate between the Méséglise Way and the Guermantes Way, the narrator alternates between a fruitless pursuit of love and an equally fruitless attempt to discover his artistic calling. This overall pattern is reiterated at a local level in discrete episodes—for example, the episode wherein Albertine has, at the narrator's urging, gone to a matinée at the Trocadéro rather than to the Verdurins. Initially delighted by the solitude that will allow him to pursue his artistic dreams, the narrator is filled with anguish when he realizes that Léa, a notorious lesbian, is on the bill of the Trocadéro, and he contrives to have the servant Françoise fetch Albertine. Once he knows that Albertine will return, however, he loses all interest in seeing her, and, while engaging in the aesthetic musings that his time with Albertine tends to preclude, he determines to devote to art his "reconquered liberty" (ML 5: 259/G 529). The subsequent flight and death of Albertine will delay his commitment to art at this time, as he suggests in a proleptic passage: "[M]y calm, and consequently the freedom that would enable me to devote myself to it, was once again to be withdrawn from me" (ML 5: 259–60/G 1751). The oscillation that takes place in this episode occurs on a global scale throughout the course of the novel as the narrator alternately desires love and literary genius. Until he has his epiphany at the Guermantes reception, the narrator will continue alternating between inaccessible goals, the theme of unfulfillment and the structure of deferral occurring at both local and global levels.

Paradoxically, this theme and structure are necessary for the text to reach the conclusion it does. Although, before the epiphany, the narrator believes that his love for Gilberte, for the Duchess de Guermantes, and for Albertine keeps him from his great work, he discovers at the reception that they actually provide "the lessons though which one serves one's apprenticeship as a man of letters" (ML 6: 316/G 2293). Through the seeming detour of the Méséglise Way, he can, in fact, arrive at Guermantes: "[T]he 'Guermantes Way,' too, on this interpretation, had emanated from 'Swann's Way' " (ML 6: 329/G 2300). At both local and global levels, the unattainable attracting points eventually reveal themselves as part of the same system. Late in life, the narrator discovers that the two walks, which had seemed so distinctive in his youth, actually meet: "Gilberte said to me: 'If you like, we might after all go out one afternoon and then we can go to Guermantes, taking the road by Méséglise, which is the nicest way,' a sentence which upset all the ideas of my childhood by informing me that the two 'ways' were not as irreconcilable as I had supposed" (ML 6: 3–4/G 2125). On the global or symbolic level, the aesthetic and erotic paths, although seemingly opposed, lead into one another. The "shape" of the narrator's desires and aspirations is like the shape of the butterfly attractor, whose one trajectory forms two wings.

Through the erotic, the narrator can arrive at the ideas that will feed into his novel, as he realizes at the Guermantes reception. His attempts to penetrate the mystery that is Gilberte or the Duchess of Guermantes or Albertine will serve as the material for the great work he undertakes to write: "And I understood that all these materials for a work of literature were simply my past life" (ML 6:304/G 2287). The suffering that the narrator experiences because of love deferred gives him access to truths otherwise unavailable:

> A woman whom we need and who makes us suffer elicits from us a whole gamut of feelings far more profound and more vital than does a man of genius who interests us. It is for us later to decide according to the plane upon which we are living, whether an infidelity through which some woman has made us suffer is of little or great account beside the truths which it has revealed to us and which the woman who exulted in our suffering would hardly have been able to understand. (ML 6: 316/G 2293–94)

Had the narrator's love for Gilberte, for the Duchess, or for Albertine been fulfilled, he would not have eventually achieved the understanding that he deems necessary for great art.

As the narrator discovers, Albertine is his best teacher: "When I was in love with Albertine, I had realised very clearly that she did not love me and I had had to resign myself to the thought that through her I could gain nothing more than the experience of what it is to suffer and to love" (ML 6: 308/G 2289). The strange-attractor structuration of the text's expanded middle enables a deepening of this theme, for the narrator's extended sufferings entail a sharper insight. In fact, the narrator's affair with Albertine, which causes him the greatest suffering, takes place in this expanded middle.[39]

As the narrative trajectory jumps between the two attracting points, it gathers up new information along the way, just as the trajectory of the strange attractor returns to certain areas, filling up state space. So, for example, the narrator discovers many years afterward that, during the time he was regularly visiting Gilberte, "she was in love with a young man of whom she saw a great deal more than of myself" (ML 5: 172/G 1703). He learns even later that she was walking in the Bois with Léa and not with a young man at all. The "trivial incident" of the syringa episode later takes on "cruel significance" (ML 5: 63/G 1643). After Albertine's death, Andrée tells the narrator that, in order to hide the signs of their lovemaking, the two women had pretended to dislike the scent of the syringa that he had brought to Albertine. When recounting the episode initially, the narrator makes an intriguing proleptic comment about his inability ever to know what really happened: "But we shall see all this—the truth of which I never ascertained—later on" (ML 5: 65/G 1644). Even when Andrée reveals new information about Albertine, the narrator cannot regard it as definitive: "Had this absence of fear permitted her to reveal the truth at last in telling me all that, or else to concoct a lie, if for some reason, she supposed me to be full of happiness and pride and wished to cause me pain?" (ML 5: 815/G 2058) There is always the possibility that new information could come to light about Albertine, about Andrée, about all the characters who play a role in *Search*—that, given time, the text will reveal "the hundred different masks which ought properly to be attached to a single face" (ML 6: 527/G 2399). Although the text is complete, its strange-attractor structuration encourages us to imagine this possibility. As with a strange attractor, new knowledge continues to be added to the overall pattern, bringing the narrator closer to the truths that will form the material of his book.

When the narrator sets out for the Guermantes reception, unfulfilled in love and unable to write, he has reached the nadir of his existence: "I knew myself to be worthless" (ML 6:239/G 2253). He continues to oscillate between the two attracting points although he begins to regard his literary aspirations as futile: "Really, I said to myself, what point is there in forgoing the pleasures of social life, if, as seems to be the case, the famous 'work' which for so long I have been hoping every day to start the next day, is something I am not, or am no longer, made for and perhaps does not correspond to any reality" (ML 6: 240/G 2253). However, at the Guermantes reception, once he realizes that the social life in which he engaged constitutes the material of his work, the oscillation can come to an end and he can dedicate himself to his writing. This conclusion recalls what may happen in a dynamical system. A dynamicist can vary the parameters in such a way that the system's behavior can be either periodic or chaotic, and a steady state attractor can become a strange attractor and vice versa (see figures 3.1 and 3.2).[40] The narrator's epiphany signifies the collapse of the strange attractor onto a fixed-point attractor. His seemingly endless travels along the Méséglise and Guermantes Ways now lead him inevitably toward the ultimate goal of Guermantes that for so long seemed unattainable. He will conclude his quest by recreating the time he "lost" as he moved toward that conclusion. Fittingly, he now meets Mlle Saint-Loup, who draws both Ways together.

In his afterword to *Narrative Discourse*, Genette makes the suggestive decree, "We must restore this work to its sense of unfulfillment, to the shiver of the indefinite, to the breath of the *imperfect*. The *Recherche* is not a closed object: it is not an object."[41] One might easily offer the following substitution: the strange attractor is not a closed object; it is not an object. By viewing Proust's text through the lens of chaos theory, we can appreciate its bounded randomness—its ability to give us a sense of a continually evolving trajectory within the closed object of the book. Through the synthesizing narrative mode of

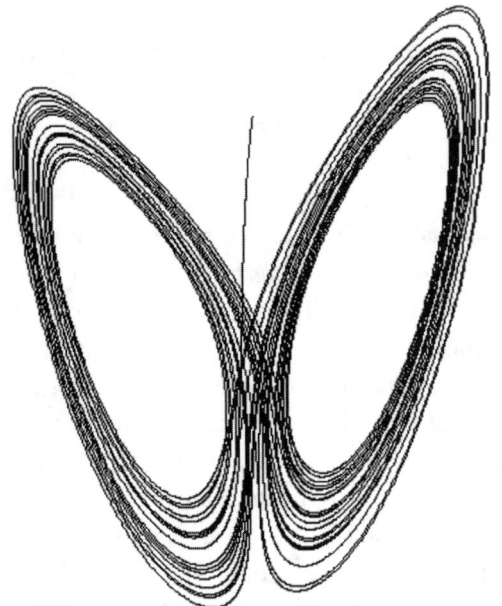

Figure 3.1 A butterfly strange attractor

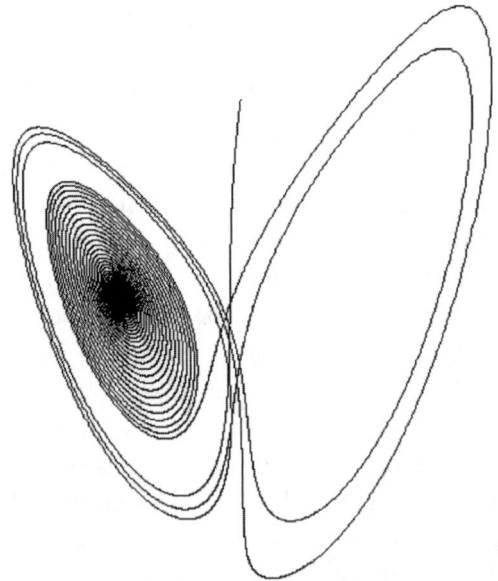

Figure 3.2 A butterfly strange attractor collapsing into a fixed-point attractor

frequency, Proust reveals "the world of potentiality" (ML 5: 21/G 1620) that is the "permanent reality" (ML5: 24/G 1622) of experience. Through a narrative trajectory that oscillates between two attracting points, he gives us the pattern of a life that can be transmuted into literature.

Ultimately, Proust's aim is didactic. The narrator claims, "every reader is, while he is reading, the reader of his own self" (ML 6: 322/G 2296). Thus although the narrator's experiences are themselves unique, through them, Proust intends us to discover our own reality: "Real life, life at last laid bare and illuminated—the only life in consequence which can be said to be really lived—is literature, and life thus defined is in a sense all the time immanent in ordinary men no less than in the artist" (ML 6: 298/G 2284).

CHAPTER 4

Narrating the Unbounded: Mrs. Dalloway's Life, Septimus's Death, and Sally's Kiss

No doubt Proust could say what I mean—
—*The Diary of Virginia Woolf*

Nature ... has further complicated her task and added to our confusion by providing not only a perfect rag-bag of odds and ends within us ... but has contrived that the whole assortment shall be lightly stitched together by a single thread. Memory is the seamstress, and a capricious one at that. Memory runs her needle in and out, up and down, hither and thither. We know not what comes next, or what follows after.

—Virginia Woolf, *Orlando*

In the midst of chaos there was shape; this eternal passing and flowing (she looked at the clouds going and the leaves shaking) was struck into stability.

—Virginia Woolf, *To the Lighthouse*

Mrs. Dalloway appears as a direct descendant of *Tristram Shandy* and *In Search of Lost Time*. The text's spiraling narrative trajectory, which works against coming to a conclusion, may remind us of the trajectory in Sterne's text. Like Sterne, Virginia Woolf appears unable to adhere to the inexorable "line of gravitation," and references to clock time, ironic counterpoints to the violations of linearity, appear throughout *Tristram Shandy* and *Mrs. Dalloway*. Like Proust's *Search*, *Mrs. Dalloway*, deals with the remembrance of things past. In it, Woolf attempts to convey "[l]ife itself, every drop of it, here, this instant, now, in the sun, in Regent's Park"—to make of it a Proustian madeline that can encapsulate a world.[1] Considering Woolf's admiration for both writers, these influences are not unexpected.[2]

Woolf found much to praise in Sterne's work, particularly *Tristram Shandy*. She appreciated Sterne's method of building up a character: "Here our sense of elasticity is increased so much that we scarcely know where we are. We lose our sense of direction. We go backwards instead of forwards. A simple statement starts a digression; we circle; we soar; we turn round; and at last we come again to Uncle Toby who has been sitting meanwhile in his black plush breeches with his pipe in his hand."[3] Woolf herself employs a similar method in creating Mrs. Dalloway—the seemingly shallow society matron whose character deepens as the narrative trajectory spirals around her. According to Woolf, *Tristram Shandy*, "in which all the usual conventions are consumed," transcended fiction: "For Sterne by the beauty of his style has let us pass beyond the range of personality into a world which is not altogether the world of fiction. It is above."[4] *Tristram Shandy*, asserts Woolf, "is complete in itself; it is self-contained," a pronouncement that sums up *Mrs. Dalloway* as well.[5]

This emphasis on creating a self-contained world appears in Woolf's assessment of Proust as well. Woolf describes his method as both expansive and transformative: "Proust, the product of the civilization which he describes, is so porous, so pliable, that we realize him only as an envelope, thin but elastic, which stretches wider and wider and serves not to enforce a view but to enclose a world. The commonest object, such as a telephone, loses its simplicity, its solidity, and becomes a part of life and transparent."[6] Certain objects in *Mrs. Dalloway*, such as the skywriting plane or Peter Walsh's penknife, similarly become "a part of life"—something more than themselves. For Woolf, Proust was "a modern with a zest," indeed "far the greatest modern novelist."[7] In many ways, *Mrs. Dalloway* serves as testimony to Woolf's admiration for both Sterne and Proust.[8]

However, although Woolf owes something to both Sterne and Proust, her version of the bounded randomness that we see in their texts is uniquely her own—her own as a woman writer. Although Woolf, in her later work *A Room of One's Own*, claimed that the act of writing was genderless, she insisted on the uniqueness of women's writing, and noted that, when a woman begins to write, she would perhaps discover "that there was no common sentence ready for her use."[9] Further, she would find that available literary forms were uncongenial as well: "a book is not made of sentences laid end to end, but of sentences built, if an image helps, into arcades or domes. And this shape too has been made by men out of their own needs for their own uses. There is no reason to think that the form of the epic or of the poetic play suits a woman any more than the sentence suits her."[10] Later in the text, while "reviewing" the fictitious Mary Carmichael's novel *Life's Adventure*, Woolf comments upon Carmichael's "tampering with the expected sequence": "First she broke the sentence; now she has broken the sequence."[11] Despite certain shortcomings, Mary Carmichael, Woolf concludes, has "mastered the first great lesson; she wrote as a woman, but as a woman that has forgotten that she is a woman."[12] It may be that the "first great lesson" that "Mary Carmichael" has mastered is one that Woolf had herself learned while writing *Jacob's Room* and *Mrs. Dalloway*.

For Woolf, writing "as a woman" has to do precisely with the breaking of the traditional, male-authorized sentence and sequence. Conversely, writing "as a man" means linear sentences and sequences, such as the unctuous Hugh Whitbread, for example, imposes upon Lady Bruton's letter: "Hugh . . . marvellously reduced Lady Bruton's tangles to sense, to grammar such as the editor of the *Times* ... must respect" (110). As Rachel Blau DuPlessis points out, men may also break sentence and sequence, but women do so because of their situation *as* women:

> There is nothing exclusively or essentially female about "the psychological sentence of the feminine gender," because writers of both sexes have used that "elastic" and "enveloping" form. But it is a "woman's sentence" because of its cultural and situational function, a dissension stating that women's minds and concerns have been neither completely nor accurately produced in literature as we know it. Breaking the sentence is a way of rupturing language and tradition sufficiently to invite a female slant, emphasis or approach.[13]

The disordering of traditional narrative structure often highlights women's concerns. Thus the structuration of *Mrs. Dalloway* is not simply a case of Woolf's putting her own spin on *Tristram Shandy* and *In Search of Lost Time* but is instead a woman writer's deliberate attempt to break sentence and sequence in order to address the thoughts and concerns of women.

In *A Room of One's Own*, Woolf acerbically comments on the critical devaluation of a woman writer's subject matter: "This is an important book, the critic assumes, because it deals with war. This is an insignificant book because it deals with the feelings of women in a drawing room. A scene in a battlefield is more important than a scene in a shop."[14] Woolf may be thinking of *Mrs. Dalloway* when she makes this point, for unsympathetic critics can indeed assess it according to this gender-based value system. As Woolf embarked upon writing the novel, she posed the following question in a letter to Gerald Brenan, a writer friend: "But how does one talk about everything in the whole of life, so that one's hair stands on end, in a drawing room?"[15] In her diary, she explained what this "everything" would entail: "I want to give life & death, sanity & insanity; I want to criticise the social system, & to show it at work, at its most intense—."[16] In defiance of what she knows will be found critically significant, Woolf chooses as a main character the party hostess Mrs. Dalloway and as the main story event a party, one in a seemingly endless succession of such glittering affairs. The shell-shocked soldier, although making an appearance on the stage, is Mrs. Dalloway's "shadow," and the Great War that leaves him psychologically maimed is referred to in passing but not described, its significance consisting in its aftereffects on people attempting to live their day-to-day lives. Breaking sentence and sequence, Woolf tells us the feelings of Clarissa Dalloway in her drawing room and sends her to a shop to buy flowers.

In assessing Woolf's achievement, we can compare *Mrs. Dalloway*'s structure to that of a fractal, which, created from a simple algorithm, displays infinite complexity. The "algorithm" for the text is simple in the extreme. Woolf prescribes a set of formal constraints for herself: she takes as her main character one who, as she herself noted, "may be too stiff, glittering, and tinsely"[17]; she restricts the actual story events to approximately seventeen hours in "real time"; and she restricts the area in which her characters interact to several largish blocks of central London, with the Dalloway drawing room as the location where the greatest amount of sustained action takes place. The present-day events that take place are fairly simple: Clarissa Dalloway prepares for a party and then gives it, while shell-shocked Septimus Smith, beleaguered by uncomprehending doctors, commits suicide. Within these global limits that she sets for herself, Woolf manages to suggest that the potential for an infinite amount of local variations exists. A strange attractor exists within a fractal dimension, and, like the dynamic of the attractor, the overall narrative trajectory of *Mrs. Dalloway* simulates ongoing evolution within a bounded area.[18] Woolf thus manages both to adhere to the constraints she has set for herself and to give us the sense that she has burst them.

The overall narrative trajectory comprises what I call the roving trajectory of focalization, which threads between the various characters in *Mrs. Dalloway*, linking them spatially, and the temporal trajectory, wherein the insistent linear time-line is consistently subverted by memories spiraling back to the past. The roving trajectory of focalization moves through a variety of "consciousnesses," enabling Woolf to establish the fluid identity of Mrs. Dalloway and suggest that she can encapsulate "everything." The temporal trajectory jumps back and forth between present and past events, often revisiting the same event from different perspectives, enabling Woolf to undermine clear causal connections, call into question a definitive truth of events, and stylistically reinforce a central theme of the novel. Although, for the sake of clarity, I look at each trajectory in isolation, they function together to effect the bounded randomness for which Woolf aimed in *Mrs. Dalloway*.

"The Whole of Life . . . in a Drawing Room"

When writing *Jacob's Room*, Woolf wrote in her diary: "Suppose one thing should open out of another . . . doesn't that give the looseness & lightness I want: doesn't that get closer & yet keep form and speed, & enclose everything, everything?"[19] This "opening of one thing out of another" is key to our understanding of what makes *Mrs. Dalloway* work. The oft-quoted passage that describes Richard Dalloway and Hugh Whitbread leaving Lady Bruton exemplifies Woolf's technique:

> And they went further and further from her, being attached to her by a thin thread (since they had lunched with her) which would stretch and stretch, get thinner and thinner as they walked across London; as if one's friends were attached to one's body, after lunching with them, by a thin thread, which (as she dozed there) became hazy with the sound of bells, striking the hour or ringing to service, as a single spider's thread is blotted with raindrops, and, burdened, sags down. (112)

The "thin thread" connecting Lady Bruton to the two men runs throughout the novel, connecting one thing to another and enabling Woolf to "enclose everything, everything."[20]

The roving trajectory of focalization constitutes this thin thread. In *Mrs. Dalloway*, Woolf shifts focalization among the consciousnesses of over forty characters. Patricia Matson argues that *Mrs. Dalloway* features a "communal protagonist": "no one point of view dominates. . . . the narrative continuously slips from one subject's point of view to another's."[21] Granted, the roving trajectory of focalization works against the dominance of a single point of view, but, Mrs. Dalloway *is* the putative protagonist of the book that bears her name, and, in order to grasp Woolf's aims, we need to turn the spotlight on the title character herself. Eponymous titles, after all, encourage such a focus and invite us to make a connection between the title character and the text itself.[22] Woolf's roving trajectory of focalization, by seeming continually to shift the focus away from Mrs. Dalloway, paradoxically establishes her identity—an identity that is fluid and unbounded rather than stable and fixed, an identity that can encapsulate the infinite within the finite.[23] The roving trajectory helps create the character Mrs. Dalloway—and the character of *Mrs. Dalloway*.

In making Mrs. Dalloway the focus of her text, Woolf has deliberately chosen a character that we might indeed dismiss as "tinsely." It is as if Woolf wished to test herself—to show that she could take what would seem to be the most superficial of creatures and imbue her with complexity and depth. In her famed response to Arthur Bennett, Woolf invented "Mrs. Brown"—"an old lady of infinite capacity and infinite variety"—who might easily be dismissed by modern writers.[24] In *Mrs. Dalloway*, she aims to show that, like Mrs. Brown, her title character is a lady of "infinite variety."

In her early incarnations, Mrs. Dalloway is particularly "tinsely." In *The Voyage Out*, she is, in protagonist Helen Vinrace's terms, "a thimble-pated creature" and "rather second-rate."[25] She is charming, snobbish, and affected, a figure of social satire. David Dowling considers Mrs. Dalloway's portrayal in *The Voyage Out* as "harsh"; her husband is "a prurient male chauvinist" and she "is not much better."[26] In the 1923 short story "Mrs. Dalloway in Bond Street" (originally intended as chapter 1 of the novel), Mrs. Dalloway has begun to take on more depth than in her first incarnation, now having a vibrant inner life and voicing opinions that do not seem mere echoes of her husband's, but she still tends to be somewhat shallow.[27] With her facile patriotism and patronizing class-consciousness, she is the stereotypical society matron.

In the novel that bears her name, Mrs. Dalloway seemingly could retain her stereotypical function. One might sum her up as the ultimate party hostess—she has a "genius" for it, as Peter Walsh acknowledges (77). Often her concerns seem superficial, and her attitudes snobbish. She herself worries that she may be "nothing—nothing at all" (10), and we may be tempted initially to agree with her self-assessment that she is a nonentity: "She had the oddest sense of being herself invisible, unseen; unknown; there being no more marrying, no more having of children now, but only this astonishing and rather solemn progress with the rest of them, up Bond Street, this being Mrs. Dalloway; not even Clarissa any more; this being Mrs. Richard Dalloway" (10–11). Marriage may no longer make women *femmes couvertes* in law, but it continues to do so in reality—in their conceptions of themselves and others' conception of them. As Peter notes, "With a mind of her own, she must always be quoting Richard. . . . These parties for example were all for him, or for her idea of him" (77). According to Peter, marriage has turned Mrs. Dalloway into a mere parrot, and even what she does best, she apparently does in service of another.[28]

Woolf strives to make her character visible, seen, and known—to fill in the missing proper name and render Mrs. Dalloway Clarissa.[29] Although giving parties may seem the quintessential superficial attainment, Clarissa through them achieves something significant, as her musings make clear:

> Here was So-and-so in South Kensington; some one up in Bayswater; and somebody else, say, in Mayfair. And she felt quite

continuously a sense of their existence; and she felt what a waste; and she felt what a pity; and she felt if only they could be brought together; so she did it. And it was an offering; to combine, to create, but to whom? An offering for the sake of offering, perhaps. Anyhow, it was her gift. (122)

Clarissa's "gift" bears affinities to that of her creator, who brings together various so-and-sos in her own text.[30] Clarissa aims to establish connections among people, just as Woolf aims to establish connections among her characters, from the lowly woman singing at the tube station to the lofty personage whose car enters Buckingham Palace.

As we move through the text, the superficial Mrs. Dalloway, fixed in her identity as party hostess, transforms into the substantial Clarissa, whose identity is fluid and indefinable. While attempting to sum up her character, Peter testifies to this substantiality: "She came into a room; she stood, as he had often seen her, in a doorway with lots of people round her. But it was Clarissa one remembered. Not that she was striking; not beautiful at all; there was nothing picturesque about her; she never said anything specially clever; there she was, however; there she was" (76). Clarissa's substantiality can be apprehended but not defined.

Woolf establishes Clarissa's substantiality through the roving trajectory of focalization, which links Clarissa's consciousness to those of the other characters in the text. Again, we are privy to the consciousnesses of over forty characters, including that of the narrator, who at certain instances functions as a character as well, engaging in imaginative projections or making didactic pronouncements. At times, the trajectory is focalized in a particular character for only a sentence or so—Scrope Purvis and Mr. Bentley at Greenwich, for example. At others, it is focalized in a character for a significant portion of the text, as with Peter Walsh, Septimus Smith, and Clarissa Dalloway herself. Sometimes it only passes through a particular character once, and sometimes it returns again and again. The trajectory of focalization may move back and forth between two characters, as in the first meeting between Peter and Clarissa, when it volleys back and forth between them like a tennis ball. It may at times seem to partake of two consciousnesses at once.[31] Woolf herself envisioned her method (her "discovery") in terms of a subterranean network: "I dig out beautiful caves behind my characters; I think that gives exactly what I want; humanity, humour, depth. The idea is that the caves shall connect, and each comes to daylight at the present moment."[32] The trajectory of focalization connects these "caves," running through the minds of the various characters.

Woolf accomplishes shifts in focalization through the contiguity of the various characters, often using a particular object (an airplane, a stumbling child, the chime of Big Ben, the play of sunlight) to effect the transition, as if the trajectory of focalization is handed off from one character to another like the baton in a relay race.[33] Although certain focalizing consciousnesses never "meet" Clarissa, they are all connected to hers through the trajectory of focalization that threads through each of them.

Consider *Mrs. Dalloway* in terms of an open dynamical system. Various consciousnesses move through that system and thereby establish the identity of the title character. A strange attractor, as Thomas Weissert points out, "takes its identity from its basin of attraction, the infinite ensemble of trajectories that at some time give up their own singular identity and take on the traces of the attractor identity."[34] The roving trajectory of focalization falls onto a strange attractor, an attractor whose shape constitutes Mrs. Dalloway—and *Mrs. Dalloway*. We apprehend Clarissa Dalloway as the sum of that infinite ensemble of other consciousnesses through which the roving trajectory moves.

Significantly, the pattern is continuously in process, or (to be more precise, as the text does come to an end) the pattern suggests infinite evolution. Mrs. Dalloway's own musings on her mortality aptly sum up what Woolf attempts to convey with Mrs. Dalloway/*Mrs. Dalloway*:

> Did it matter then, she asked herself, walking toward Bond Street, did it matter that she must inevitably cease completely; all

this must go on without her; did she resent it; or did it not become consoling to believe that death ended absolutely? but that somehow in the streets of London, on the ebb and flow of things, here, there, she survived, Peter survived, lived in each other, she being part, she was positive, of the trees at home; of the house there, ugly rambling all to bits and pieces as it was; part of people she had never met; being laid out like a mist between the people she knew best, who lifted her on their branches as she had seen the trees lift the mist, but it spread ever so far, her life, herself. (9)

Although Mrs. Dalloway may die (but, provocatively, does not in this text), although the book that holds her life between its covers may come to an end, Woolf implies that she as a writer could go on indefinitely, creating new characters, some of whom may not even meet Clarissa Dalloway and all of whom would contribute to the continually emerging attractor that constitutes her being—"being" as participle rather than noun.

The most important contributor to the "emerging attractor" Clarissa Dalloway is Septimus Smith, a character whose actual spatial path never crosses hers but with whom she feels a sympathetic resonance. In her initial conception of the novel, Woolf intended for Clarissa to die, perhaps to kill herself. Yet some time during its composition, Woolf added Septimus and abandoned her earlier plan. Septimus allows Woolf to deal with "life & death, sanity & insanity" in a way she might not otherwise have been able to do. In her diary, she noted that it is Septimus who enables her to turn her story of Clarissa Dalloway into a book: "Mrs. Dalloway has branched into a book; and I adumbrate here a study of insanity & suicide: the world seen by the sane and the insane side by side—something like that. Septimus Smith?—is that a good name?"[35] In a letter to Brenan, she spoke of the interdependency of Septimus and Clarissa: "Septimus and Mrs. Dalloway should be entirely dependent upon each other—if as you say he 'has no function' in the book then of course it is a failure."[36] Septimus's insanity becomes a part of the ostensibly normal Clarissa, Woolf thus suggesting that all of us are susceptible to a dual vision of the world—sane and insane. Notably, after seemingly experiencing his death, Clarissa ponders her own madness: "[T]here was in the depths of her heart an awful fear. Even now, quite often if Richard had not been there reading the *Times*, so that she could crouch like a bird and gradually revive, send roaring up that immeasurable delight, rubbing stick to stick, one thing with another, she must have perished" (185). Septimus's wartime experiences become a part of *Mrs. Dalloway*, enabling Woolf to show that what happens on the battlefield impacts upon what occurs in the drawing room and that the separation of male and female realms is a specious invention by the critics. The disparate, disordered fragments of contemporary life merge in an overall pattern.

Most important, Septimus's death ends up being absorbed into Clarissa Dalloway's life. The last incursion into Septimus's consciousness ends with him flinging "himself vigorously, violently down on to Mrs. Filmer's area railings" (149). Interestingly, after Clarissa is told of his death at her party, her consciousness takes up where his left off when he jumped out the window: "Up had flashed the ground; through him, blundering, bruising, went the rusty spikes. There he lay with a thud, thud, thud in his brain, and then a suffocation of blackness" (184). Woolf tells us, "So she saw it" (184), but Clarissa has felt Septimus's death as well, his death essentially serving as a surrogate for her own. Had she herself indeed died, as Woolf had initially conceived, we would have ended up with a finished portrait of Clarissa, the meaning of her life summed up by her death. As DuPlessis points out, killing off a heroine, like marrying her off, adheres to the "cultural practice of romance": "Marriage celebrates the ability to negotiate with sexuality and kinship; death is caused by inabilities or improprieties in this negotiation, a way of deflecting attention from man-made social norms to cosmic sanctions."[37] The creation of Clarissa's shadow aids Woolf in her attempt to establish Mrs. Dalloway's life as a life-in-process, one that, like the mist, will continue to "spread ever so far," or, like the strange attractor, will continue to evolve—bounded but not by the traditional narrative constraints of marriage or death that have tended to define female existence.

This attempt to convey a life-in-process may explain in part Woolf's return to Mrs. Dalloway's party in the half-dozen stories she wrote after completing her novel. Stella McNichol, who has compiled these six stories and the earlier "Mrs. Dalloway in Bond Street" in *Mrs. Dalloway's Party*, notes, "It is particularly uncharacteristic of Virginia Woolf's normal writing habits that she should have allowed her completed novel's central concern to retain the hold on her imagination that it obviously then had." McNichol provides a couple of explanations for the stories' exact relationship to the finished novel; they may have been envisioned "either as parts of the novel itself, later to be rejected and to swim free as independent stories, or as alternative parallel expressions of Virginia Woolf's ideas."[38] Certainly, they do "swim free" of *Mrs. Dalloway*, yet we can also regard them as Woolf's effort to show that "on the ebb and flow of things," Mrs. Dalloway "survived." Although Mrs. Dalloway appears *in propria persona* in only three stories of the six, and even then but briefly, we can imagine the thoughts of such characters as Prickett Ellis or Lily Everit being incorporated into the trajectory of focalization that spirals through *Mrs. Dalloway*, just as the thoughts of Scrope Purvis are. We may even imagine an alternative history with Woolf as a latter-day Sterne, unable to bring her text to a conclusion and continuing to spin out new stories that have some connection, however tangential, with Mrs. Dalloway and her party.

The strange attractor that emerges as we follow the pattern of the roving trajectory enables Woolf to put forward the feminist value of connectedness. Patricia Matson notes that in embracing "the wonders of multiplicity," Clarissa "defies the exclusive/excluding codes of phallogocentric discourse,"[39] and these values are those of Woolf as well. In a particularly poignant passage, Clarissa ruminates upon the essential separateness of the human condition: "here was one room; there another. Did religion solve that, or love?" (127). In a world that had recently been riven by the divisive act of war, Woolf attempts to solve the problem of the two rooms—attempts to establish the value of connectedness linked (although not essentially) to the female realm—through the roving trajectory of focalization that brings together in an ordered pattern the chaos of separate lives.

"If It Were Now to Die . . ."

For its narrative complexity *Mrs. Dalloway* depends not merely upon Woolf's innovative and peculiar use of the roving trajectory of focalization. It depends also on scrambling of chronology. Although Woolf subverts the linear time-line, she reorders events according to a chaotic attractor structure. This "breaking of sequence," enables Woolf concurrently to criticize patriarchal structures and authority and to dissemble her critique—to avoid the either/or of conclusiveness.

Throughout the novel, Woolf emphatically enforces time's linear progression as she resists it. The present-day story events of *Mrs. Dalloway* take place over the course of seventeen hours, what we might consider the typical time that we spend awake. The bongs of Big Ben, powerful reminders of the inexorable passage of time, regularly punctuate the action. However, for Woolf, as for Sterne, this linear progression is something to be feared. Clock-time tolls mortality, as Clarissa's musings after Lady Bruton has slighted her, suggest: "But she feared time itself, and read on Lady Bruton's face, as if it had been a dial cut in impassive stone, the dwindling of life; how year by year her share was sliced." (30). When Peter listens to the clock stroke from St. Margaret's, he connects it with Clarissa's illness: "It was her heart, he remembered; and the sudden loudness of the final stroke tolled for death that surprised in the midst of life, Clarissa falling where she stood, in her drawing-room" (50). Like Sterne, Woolf equates the march of time with death.

Woolf explicitly links clock time with a masculinist system of values—with authoritarianism, with the imposition of one's will upon another, with, in fact, the patriarchal will to power exemplified by Sir William Bradshaw. Big Ben, the exemplar and enforcer of linear time in the novel, "represents the Father," according to Makiko Minow-Pinkey, in that "it dissects the continuum of life and imposes a structure."[40] In an extended disquisition, the narrator tells us that Sir William is a devotee of the Goddess Proportion and her "less smiling more formidable" sister Conversion, who "feasts on the wills of the weakly, loving to impress, to impose" (100). Significantly, the disquisition is followed first by Rezia's cry that she does not "like that man" and then by a description of the city clocks: "Shredding and slicing, dividing and subdividing, the clocks of Harley Street nibbled at the June day, counselled submission, upheld authority, and pointed out in chorus the supreme advantages of a sense of proportion." (102). Through her adherence to a rigid time scheme in recounting the events of one day in June, Woolf pays lip service to the death-dealing, patriarchal law of the clock, but through the particular narrative structuration she employs, she undermines it.

For, despite the emphasis on the time-line, the temporal trajectory of the plot is not linear at all. As it moves forward in the present day, it continually jumps back to earlier times, covering a segment of the past here, a segment of the past there, but never filling in all the gaps. As in Sterne and Proust, memory is the vehicle that takes characters and readers into the past. We flash back to Clarissa Dalloway's courtship summer at Bourton and Septimus Smith's wartime experiences. At least once, we even flash forward—when we hear of the probable sundering of Peter Walsh's relationship with his common Daisy.

Woolf not only jumps around in time, but she also the blurs the boundary between past and present. Woolf accomplishes this "ambiguity of the temporal location" through her use of the simple past for both the narrator's commentary and the character's memory.[41] A fairly substantial flashback detailing the final break between Clarissa and Peter, for example, is followed by the lines: "It was awful, he cried, awful, awful!" (64). The emotion invoked by the lines would seem still to be part of

Peter's past, but the subsequent lines take us into the present: "Still, the sun was hot. Still, one got over things." The overall effect of this technique shows the difficulty of disentangling past events and present states, thus aiding Woolf in her subversion of linear temporality.

Like the evolution of the strange attractor, the evolution of the temporal trajectory manifests a disorderly order. In order to visualize this evolution, we might think of the novel in terms of a temporal grid. Along one axis, we can chart the story events, both past and present, which take place over an approximately thirty-year period. Along the other axis, we can chart the time of our reading—or, more accurately, our movement through the text itself. The resultant trajectory, we find, follows deterministic laws—it will move through seventeen hours of June 13—but the particular way in which it will do so is unpredictable.

As it moves back and forth between past and present, the temporal trajectory falls into a certain pattern. It completely bypasses certain areas on the temporal grid—for example, many of the years between Clarissa's marriage and the present-day events. It passes only briefly through certain segments of the past, as when Clarissa recalls her "failure" at Constantinople or when Mrs. Dempster recalls her time at Margate. At times, it revisits a particular area on the temporal grid again and again, such as the courtship summer at Bourton and the war years, these particular areas on the grid constituting basins of attraction. When revisiting a particular area, the temporal trajectory may retrace the events from the perspectives of two or more different characters. For example, the breakup between Peter and Clarissa is dealt with from Peter's, Clarissa's, and Sally's perspectives, and the memories of each cover slightly different, although overlapping, segments on the temporal grid. As with the trajectory of focalization, the temporal trajectory simulates a certain infinite evolution. It could visit other areas on the temporal grid or return to those it had already visited. Like the trajectory of the strange attractor, it comes to no conclusion.

The nonlinear structuration enables Woolf both to criticize her society's constricting gender norms in the story events themselves and to avoid conclusive pronouncements. The skewing of linear chronology undermines a clear cause-effect sequence; the revisiting of a past event from different perspectives and the seemingly continuous evolution of the temporal trajectory undermine a definitive statement of truth. Although the movement of the temporal trajectory to particular basins of attraction gives us clues to the impact of past events upon present conditions, Woolf leaves it up to us readers to discern an emergent pattern of the text's dynamical system and to fill in gaps that she deliberately leaves open.[42] Once we discern the pattern, we can piece together two prequels from the various flashbacks, which, as is typical for the work of memory, appear in no particular order. We can then attempt to reconstruct a plot for each of these prequels, to forge a causal chain. If we concentrate on the backstories of Clarissa and Septimus, the causal chain that we forge enables us to discern the depredations wreaked on the psyche when one submits to prescribed gender roles.

Clarissa's backstory includes her suppression of a lesbian attachment in order to take her place in society that determines a woman's role to be that of a wife and mother. Jane Marcus succinctly sums up her character, "She is the lesbian who marries for safety and appearances, produces a child, cannot relate sexually to her husband, and chooses celibacy within marriage, no sex rather than the kind she wants."[43] Although Peter Walsh and Richard Dalloway see themselves as rivals for Clarissa's hand during the summer at Bourton, the true object of her affections, which neither is able to discern, is the hoydenish Sally Seton. Significantly, the temporal trajectory returns several times to Sally's naked run down the corridor at Bourton, attracted to this site of female eroticism. When Clarissa recalls what she felt when Sally was "beneath this roof" during the summer at Bourton, the description is similar to what we might expect to read in a traditional story of heterosexual romance:

> But she could remember going cold with excitement, and doing her hair in a kind of ecstasy (now the old feeling began to come back to her, as she took out her hairpins, laid them on the dressing-table, began to do her hair), with the rooks flaunting up and down in the pink evening light, and dressing, and going downstairs, and feeling as she crossed the hall "if it were now to die 'twere now to be most happy." That was her feeling—Othello's feeling, and she felt it, she was convinced, as strongly as Shakespeare meant Othello to feel it, all because she was coming down to dinner in a white frock to meet Sally Seton! (34–35)

Certainly, she cherishes no such Shakespearean feelings for her suitors Peter and Richard. They play no part in "the most exquisite moment of her whole life": "Sally stopped; picked a flower; kissed her on the lips" (35).

Othello's exultant statement, we recall, marks the last instance of his happiness, and the "exquisite moment" of the kiss is apparently the only one that Clarissa and Sally share—a marker for a road that is too dangerous to travel. We hear of no other physical intimacy between the two women, and during the courtship summer, Sally actually seems to be aiding and abetting Peter's cause—writing him long letters when he is away, praising him to Clarissa, "sweeping him off for talks in the vegetable garden" (63), and imploring him, "half laughing of course, to carry off Clarissa, to save her from the Hughs and Dalloways and all the other 'perfect gentlemen' who would 'stifle her soul' " (75). Each woman ultimately chooses to submit to the gender expectations of the time. Sally becomes Lady Rosseter and the mother of "five enormous boys" (171). Clarissa marries Richard Dalloway and produces the lovely Elizabeth. Once each woman marries, the relationship between them dwindles. As Sally tells Peter, "[T]he Dalloways had never been once" to visit the Rosseters (190). Sally ascribes this neglect to Clarissa's snobbery, but we must wonder if it is also due to an attempt on Clarissa's part to cut herself off completely from a dangerous liaison. Would Clarissa have preferred that Sally, rather than marrying "a bald man with a large buttonhole," had met the alternative fate that she had once predicted for her: "[I]t was bound, Clarissa used to think, to end in some awful tragedy; her death; her martyrdom" (182)? When Sally shows up uninvited at Clarissa's party, she waits in vain for Clarissa to talk to her. As readers, we may in fact expect and desire such a scene—a scene that would enable the two to address in the present the crucial event of the past. But whatever revelation that either has in seeing the other occurs separately, Woolf thereby suggesting that, even after all this time has elapsed (or perhaps because it has), what happened between Clarissa and Sally that long-ago summer cannot be openly discussed.

By marrying Richard, Clarissa has indeed played it safe. She clearly has a stronger connection with Peter than with Richard: "They went in and out of each other's minds without any effort," Peter recalls (63). Yet the very strength of this connection causes her to choose marriage with the latter: "For in marriage a little licence, a little independence there must be between people living together day and day out in the same house; which Richard gave her, and she him . . . But with Peter everything had to be shared; everything gone into" (8). Peter wishes to penetrate both mind and body, the penknife with which he constantly fiddles serving as Woolf's playful symbol of his thwarted sexual urge. Richard, however, allows Clarissa to "sleep undisturbed" in her "narrow" bed, her "virginity preserved through childbirth" (31).

Clarissa makes a marriage of convenience that enables her to fulfill her traditional womanly role while maintaining quasi-celibacy, but the psychic cost is high. She regards her lack of passion for Richard as a sign of failure: "[T]hrough some contraction of this cold spirit, she had failed him" (31).[44] She can "dimly perceive" what passion might be—"something warm which broke up surfaces and rippled the cold contact of man and women, or of women together" (31). Woolf suggests in imagery reminiscent of female orgasm that Clarissa has felt such passion momentarily when "yielding to the charm of a woman":

> It was a sudden revelation, a tinge like a blush which one tried to check and then, as it spread, one yielded to its expansion, and rushed to the farthest verge and there quivered and felt the world come closer, swollen with some astonishing significance, some pressure of rapture, which split its thin skin and gushed and poured with an extraordinary alleviation over the cracks and sores. (32)

Yet such moments are brief, and against them "there contrasted . . . the bed and Baron Marbot and the candle half-burnt" (32)—signs of a constricted and unfulfilled life. Clarissa has deliberately chosen this life: "She had wanted success. Lady Bexborough and the rest of it" (185). Certainly, the success represented by Lady Bexborough could not have been attained had she given in to her lesbian desires. But Clarissa calls up a poignant image of the night of Sally's kiss as a counterpoint to this success: "And once she had walked on the terrace at Bourton" (185). Submission to society's expectations entails loss.

Septimus's backstory, too, tells of submission. Just as Clarissa learns to submit to society's expectations about what it means to be a proper lady, the idealistic, poetry-writing Septimus learns what it means to be a proper man. The Great War serves as his teacher: "There in the trenches the change which Mr. Brewer desired when he advised football was produced instantly; he developed manliness." (86). Manliness entails the suppression of emotion—in Septimus's case, what might be a quite natural expression of grief at the death of his friend Evans: "[W]hen Evans was killed, just before the Armistice, in Italy, Septimus, far from showing any emotion or recognising that here was the end of a friendship, congratulated himself on feeling very little and very reasonably. The War had taught him. It was sublime" (86). Like the speaker in Coleridge's "Dejection: An Ode," who steals from himself "all the natural man," Septimus, once peace comes, finds "that he could not feel" (86).

Although Woolf is somewhat ambiguous about the exact nature of Septimus's relationship to Evans, Eileen Barrett makes a compelling argument that his feelings for Evans are homosexual and that his suicide serves as Woolf's condemnation of "her culture's silencing of homosexuality and its insistence on heterosexuality."[45] If we regard Septimus as cherishing a homosexual passion for Evans, we see even stronger parallels than we might otherwise see between his story and Clarissa's. But whether we regard Septimus as a silenced homosexual or simply a shell-shocked veteran, we can assume that his plight is due to his having tried to fit himself into a role of manly behavior that goes against who he is. Enjoined to eat porridge and take up a hobby by Dr. Holmes, to bow to the goddess Proportion by Sir William, Septimus ultimately opts not to submit, but he can see no other way to escape the forces of authority that surround him than to commit suicide.

When she has Clarissa seem to experience Septimus's death, Woolf drives home the theme of submission. Clarissa contrasts Septimus, who "had flung it away," with herself and her friends, who "would grow old," and she regards him as having retained an integrity that they had not: "A thing there was that mattered; a thing, wreathed about with chatter, defaced, obscured in her own life, let drop every day in corruption, lies, chatter. This he had preserved" (184). For Clarissa, what he has preserved connects vitally with the feelings that she once had for Sally: "But this young man who had killed himself—had he plunged holding his treasure? 'If it were now to die, 'twere now to be most happy,' she had said to herself once, coming down in white" (184). The temporal trajectory here revisits the site of Clarissa's greatest happiness—the time at Bourton with Sally. Clarissa's "treasure," Woolf suggests, was lost to her when she submitted herself to the heterosexual norm governing her society.

In setting up the backstories of Clarissa and Septimus, Woolf allows other versions of past events to come into play as well, for the temporal trajectory revisits certain areas but from different perspectives. No single character has access to all the memories and all the pieces of information that would explain the present state of affairs in light of past events, and Peter's version of the past and Sally's and Rezia's are as much as part of the overall pattern as Clarissa's and Septimus's. Thus, from one perspective, the novel explores thwarted homosexual desire, but, from another, it deals with thwarted heterosexual desire. It is no wonder that an early reviewer could assert that the novel's "sole principle event is the return from India of Mrs. Dalloway's rejected suitor."[46]

In *A Room of One's Own*, Woolf asserted that "when a subject is highly controversial—and any question about sex is that—one cannot hope to tell the truth."[47] Woolf's motifs of same-sex love and critique of gender roles are obscured in part by the particular structuration of *Mrs. Dalloway*, which moves toward and away from unambiguous meanings and definitive statements. Neither motif has been a given in critical assessments of the novel.[48] Considering the hostility to nontraditional gender roles and to lesbian issues during the time in which Woolf wrote, such a structuration serves a pragmatic, as well as aesthetic purpose.[49] The strange-attractor pattern works against clear causality and against an authoritative version of events, enabling Woolf to mask her critique.

Furthermore, the chaotic evolution of the temporal trajectory works against the text coming to a conclusion that would give definitive meaning to all that has gone before. Clarissa's reliving of Septimus's death might be regarded as a climax of sorts in that she appears to come to a realization about the emptiness of the choice that she has made. We are left, however, with no definitive answer about the consequences of that realization.[50] Peter may feel "terror" and "ecstasy" when he sees Clarissa coming toward him (194), but Woolf leaves unanswered what ensues when they actually meet. The final line of the text—"For there she was" (194)—might serve as the culminating moment in a traditional story of heterosexual romance, but here it serves no such function. As far as we know, Clarissa never actually talks to Sally. We are left with the sense of indefinite deferral, the sense that the temporal trajectory could continue to evolve, filling in more and more of a pattern that will never be fixed.[51] As Woolf reached her "last lap" in writing *Mrs. Dalloway*, she asked in her diary whether she could "keep the quality of a sketch in a finished & composed work?"[52] In some sense, the apparently infinite evolution of the attractor structure allows Woolf to achieve that aim.

Ultimately, the roving trajectory of focalization and the spiraling trajectory of temporal order cannot be dealt with in isolation. As we move from character to character, we are moving throughout the temporal grid. The spatial and temporal are intertwined, as in the strange attractor, whose temporal evolution is charted in state space. Like the strange attractor, *Mrs. Dalloway*, in its very boundedness, makes a good approximation of showing us "everything, everything!"

Our experience of *Mrs. Dalloway* overall is analogous to our experience of the character who gives the book its name. We cannot definitively "know" Clarissa Dalloway. She may be the sum of her parts, but the variables that constitute her are infinite and incalculable: her memories of the past and her actions in the present; her thoughts of Bourton and Septimus's thoughts of Evans; Peter Walsh's pronouncements upon her character and Scrope Purvis's upon her looks; the narrator's discussion of Proportion and Conversion and description of Miss Kilman greedily gulping "the last inches of the chocolate éclair" (132); and so on. Woolf makes clear that we cannot "say of any one in the world now that they were this or were that" (8). Clarissa herself is neither the would-be lesbian lover of Sally Seton nor the former flame of Peter Walsh, but both and more. Peter cannot sum her up; he can only say, "For there she was" (194). We can no more sum up *Mrs. Dalloway* than Peter can Clarissa. When, after pondering Septimus's death, Clarissa Dalloway thinks that, before returning to her party, "She must assemble" (186), her words suggest what Woolf prompts us to do—not only assemble Clarissa but also assemble the text itself from those infinite and incalculable variables. In doing so, we discover the emergent pattern of meaning.

Woolf set herself two problems when writing *Mrs. Dalloway*, problems that merged aesthetic and ideological concerns. She wished to encapsulate the infinite with the finite, to convey the boundlessness and connectedness of all things—what she saw as a female way of knowing the world. She also wished to address the controversial subjects of constricting gender roles and same-sex love while avoiding the definite pronouncements that could arouse hostility in an unreceptive audience. By adopting a narrative structuration that bears affinities to the strange attractor, she could solve both problems, giving us the simple but complex *Mrs. Dalloway.*

CHAPTER 5

Narrating the Indeterminate: Shreve McCannon in *Absalom, Absalom!*

> Maybe nothing ever happens once and is finished. Maybe happen is never once but like ripples maybe on water after the pebble sinks, the ripples moving on, spreading, the pool attached by a narrow umbilical water-cord to the next pool which the first pool feeds, has fed, did feed, let this second pool contain a different temperature of water, a different molecularity of having seen, felt, remembered, reflect in a different tone the infinite unchanging sky, it doesn't matter: that pebble's watery echo whose fall it did not even see moves across its surface too at the original ripple space to the old ineradicable rhythm.
>
> —Quentin Compson in *Absalom, Absalom!*

> *But now we discover an abundance of systems whose behaviour, although governed by precisely-known laws, cannot be predicted even in principle because they are so unstable. To know the law is not necessarily to know the behaviour.*
>
> —Michael Berry, "Chaology: The Emerging Science of Unpredictability"

Like *Mrs. Dalloway*, William Faulkner's *Absalom, Absalom!* deals with a minimal number of plot events. One hot afternoon, Quentin Compson listens to Miss Rosa Coldfield tell the story of the Sutpens; that evening he listens to his father fill in gaps in Miss Rosa's story"; and, six months later at Harvard, he and Shreve McCannon stay up all night piecing together the rest of the puzzle. The only events taking place in the present are instances of storytelling, and the one present-day action scene—Quentin's journey with Rosa to Sutpen's Hundred—comes to us only through a flashback. Like Virginia Woolf, however, William Faulkner attempts to convey "everything, everything!" within certain constraints. Instead of a drawing on a roving trajectory of focalization, he focalizes the overall narrative through various internal narrators in turn, and each narrator contributes to the emerging pattern of the Sutpen story.

In a 1983 discussion of *Absalom, Absalom!*, Hugh Ruppersberg pointed out that interpretations of the novel "usually fall into the 'Detective' or 'Impressionist' schools of criticism," either emphasizing the narrative structure at the expense of the story of the Sutpen dynasty or emphasizing the story at the expense of the structure.[1] In the decades that followed, the privileging of one emphasis over another has persisted, in part because of the nature of the text itself. On the one hand, the story of Sutpen is compelling in its own right: a poor white who, by dint of a monomaniacal desire to found a dynasty, becomes a wealthy, slave-owning plantation owner, facilitates the murder of his unacknowledged son of mixed-race to keep his daughter from marrying that son, and eventually destroys all that he holds dear. On the other hand, when we consider how the story comes to us—chronologically disordered and filtered through multiple, often unreliable, perspectives—we are invited to focus on how the text self-reflexively comments on the process that presumably brings it into being. Essentially all talk and no action, *Absalom, Absalom!* may simply be regarded as a narrative about narrative—a representation of representations, a ripple without an originary pebble, an echo without an originary sound.

The latter focus is amenable to a post-structuralist critical perspective that argues for the infinite deferral of meaning inherent in all discourse. In *Reading for the Plot*, Peter Brooks puts forth what is perhaps one of the most cogent explorations of the way in which *Absalom, Absalom!* signifies its own textuality. Brooks subsumes Sutpen's story under the plot of its narration, suggesting that there may be no there there:

> A further, more radical implication might be that the implied occurrences or events of the story (in the sense of *fabula*) are merely a by-product of the needs of plot, indeed of plotting, of the rhetoric of the *sjužet*: that one need no longer worry about the "double logic" of narrative since event is merely a necessary illusion that enables the interpretive narrative discourse to go further. . . . This in turn might imply that the ultimate subject of any narrative is its narrating.[2]

Although he touches on the issue of miscegenation, Brooks is more concerned with its semiotic than representational implications, arguing that it signifies a "'wild,' uncontrollable metonymy." Brooks's concentration on the text's palindromic structure and the reader's "assumption of complete responsibility for the narrative" ultimately makes a case for a "centerless" narrative.[3]

Stimulating as Brooks's study is, its focus effectually divorces the narrative from any connection with the "historical" circumstance of miscegenation leading to the fall of the house of Sutpen. Post-structuralist narratology does little to illuminate Faulkner's thematic indictment of slavery. As Ian Mackenzie has pointed out in his essay "Narratology and Thematics," narratives "*are* always *by* somebody and *about* something," and narratological operations "cannot adequately deal with the thematic interest that generally inspires our acts of responding to narratives."[4] Susan Sniader Lanser decries a narratological concentration on the "specific, semiotic, and technical" at the expense of the "general, mimetic, and political."[5] The downplaying of theme in some critical evaluations of *Absalom, Absalom!* tends to deny the importance of the slavery issue, much as Thomas Sutpen denies his own issue in the person of the mixed-race Charles Bon.

However, by employing a chaos-theory model, we can connect the text's multiperspectival, looping narrative structure—seemingly circling round a hollow center—with the theme of the South's denial of the black blood upon which it is built.[6] Radically nonlinear, Faulkner's text provides a fruitful area for exploring the way in which a meaning structure emerges out of an apparently chaotic flux. Thomas Sutpen's history demonstrates the butterfly effect common to chaotic structures—the nonlinear development of small causes into great effects or sensitive dependence on initial conditions. The absent center of a reading such as Brooks's can more appropriately be characterized as a strange attractor. Whereas applying a post-structuralist methodology to *Absalom, Absalom!* leads us into the hall-of-mirrors view that there is nothing but infinitely proliferating textuality, chaos theory enables us to discuss the text in terms of an infinite play of signification that operates within the bounded arena of a strange attractor. A chaos-theory reading thus provides a heuristic framework for modeling the narrative dynamics of *Absalom, Absalom!* that brings together its metanarrative structure and theme, while avoiding claims of the text's radical indeterminacy as well as claims of its ultimate meaning.

Thomas Sutpen and Classical Determinism

The logical extension of Newtonian deterministic thinking is Laplace's demon, that imaginary entity that can envision all past and future states of the universe. As James Crutchfield et al. point out, however, such thinking undermined notions of free will: "The literal application of Laplace's dictum to human behavior led to the philosophical conclusion that human behavior was completely predetermined: free will did not exist."[7] If we reside in a completely predictable universe, its human components are as subject to deterministic laws as anything else in the system. Chaos theory, as we know, shoots holes in the notion that all can be predicted in the macroscopic natural world and, by extension, in human behavior. Although systems can be deterministic, we may not be able to predict their future states—certainly not the future state of the extremely complex system that constitutes human behavior.

With his portrait of Thomas Sutpen—a "demon" in his own right, according to Miss Rosa Coldfield—Faulkner explores the consequences of deterministic thinking as it applies to human behavior. Sutpen's entire adult career is based on fulfilling his design. As Quentin puts it, Sutpen regards himself as a sort of baker, who needs only to mix the right ingredients together to achieve the desired result: "[I]t was that innocence again, that innocence which believed that the ingredients of morality were like the ingredients of pie or cake and once you had measured them and balanced them and mixed them and put them into the oven it was all finished and nothing but pie or cake could come out."[8] When one follows a recipe, even making a few slight variations, a pie or a cake results—over and over and over again. Sutpen expects human behavior to be deterministic and determinable. However, as Faulkner indicates, human behavior defies predictability.

By conceiving his design as a classically deterministic system, Sutpen, the tragic counterpart of Walter Shandy, brings about his own undoing.[9] The implementation of his design exemplifies the butterfly effect—namely, little causes bring about big results. The undelivered and clearly inconsequential message to the rich Pettibone—"*so he cant even know that Pap sent him any message and so whether he got it or not cant even matter, not even to Pap*" (192)—and the barring of Pettinbone's front door, a no doubt a regular and unremarkable occurrence in Pettibone's household, lead to the dirt-poor Sutpen wresting a plantation out of the land and setting up his progeny, briefly and fatally, as aristocrats of the Old South.

Caught up in deterministic thinking, Sutpen cannot account for the unpredictable—although he himself represents an unpredictable element at the local level within the global stability of the Southern caste system. In one of the few passages in

Sutpen's own words (if we can so designate an account that comes through the intervening "consciousnesses" of General Compson, Mr. Compson, and Quentin), Sutpen puzzles over the failure of his meticulously planned design: "You see I had a design in my mind. Whether it was a good or bad design is beside the point; the question is, Where did I make the mistake in it, what did I do or misdo in it, whom or what injure by it *to the extent which this would indicate.* I had a design" (212; my emphasis). What leads to the destruction of the design is the random element that causes the system trajectory to diverge from its predicted course—in this case, the fact of Eulalia Bon's black blood, an amount apparently so negligible as to allow her to pass herself off successfully as half-Spanish. A paradoxical situation occurs here: Sutpen knows that this negligible amount cannot be encompassed into the design of the patrilineal South (an implicit acknowledgment that deterministic thinking can only maintain its sway by suppressing perturbations), yet he thinks that the denial of this small perturbation in his own past will have no future consequences.

The metaphor Quentin uses when discussing Sutpen's account of his first wife reinforces our sense of unpredictable, destabilizing forces at work: "He also told Grandfather, dropped this into the telling as you might flick the joker out of a pack of fresh cards without being able to remember later whether you had removed the joker or not, that the old man's wife had been a Spaniard." (203). The "Spanish" blood is the joker in the pack—that wild card whose presence may disrupt the most winning system. By countenancing it, Sutpen would "see [his] design complete itself quite normally and naturally and successfully to the public eye," yet he would regard its culmination as "a mockery and betrayal of that little boy who approached that door fifty years ago and was turned away" (220). Allowing the possibility of miscegenation and incest within the family structure works against Sutpen's fixed idea of the flawless dynasty that he wishes to erect in response to Pettibone's rebuff.[10]

Sutpen clings to his design, his deterministic thinking, in the face of recurring destabilizing perturbations. Seemingly small causes lead to the catastrophe at Sutpen's Hundred. Charles Bon's black blood (half that of Eulalia and so insignificant as to be indiscernible) leads to Sutpen's rejection of Bon, and "that flash, that instant of indisputable recognition" (255) that he withholds from Bon sets Bon on his destructive course to claim Judith. Sutpen's spur-of-the moment words to his fiancée Rosa Coldfield cause her to flee the plantation and immure herself in the Coldfield home, bitterly nursing the insult. His casually brutal repudiation of Milly Jones leads Wash Jones to kill him. A thinking that cannot account for random elements and non-linear effects is doomed to failure.

Significantly, Sutpen's desire to become a wealthy plantation owner "with dressed-up niggers," who can "lie in a hammock all afternoon with his shoes off" (185), only replicates an already-existing social design, one that is itself predicated upon (patri)linear thinking. By clinging to that design, Sutpen actually undermines the very reason for its inception. He initially plans not only to be the wealthy plantation owner, but also to change the terms by which such a man exerts his power:

> [H]e would take that boy in where he would never again need to stand on the outside of a white door and knock at it: and not at all for mere shelter but so that that boy, that whatever nameless stranger, could shut that door himself forever behind him on all that he had ever known, and look ahead along the still undivulged light rays in which his descendants who might not even hear his (the boy's) name, waited to be born without even having to know that they had once been riven forever free from brutehood just as his own (Sutpen's) children were—. (210)

By thinking deterministically, however, Sutpen can neither replicate the preexistent social design nor rework the pattern. It is the nameless stranger/his first-born son whom he turns away from the "white door," to whom he sends, in Bon's words, "a message like you send a command by a nigger servant to a beggar or a tramp to clear out" (272)—a message such as Pettibone sent him. Sutpen's great-grandson, the mentally deficient Jim Bond, disappears without a trace, presumably fallen into the "brutehood" from which Sutpen "had once been riven." The pattern is, or course, reworked, but in a manner Sutpen had not intended, and what he replicates are the evils he had hoped to change.

Lest we not get the point about the flaws of classical determinism, Faulkner underscores it with his portrait of the conniving lawyer, whose balance sheet, as has often been remarked, parodies Sutpen's design. He attempts to equate mathematically what he can actually gain from Sutpen, even factoring the possible incest threat into the accounting: "*Incest threat: Credible Yes* and the hand going back before it put down the period, lining out the *Credible*, writing in *Certain*, underlining it" (248). Here the moral consequences of deterministic, linear thinking come into play—acquisitiveness, the urge to dominate and control.[11] But, like Sutpen, the lawyer too cannot account for the unpredictable:

> [H]e was never worried about what Bon would do when he found out; he had probably a long time ago paid Bon that compliment of thinking that even if he was too dull or too indolent to suspect or find out about his father himself, he wasn't fool enough not to be able to take advantage of it once somebody showed him the proper move; maybe if the thought had ever occurred to him that because of love or honor or anything else under heaven or jurisprudence either, Bon would not, would refuse to, he (the lawyer) would even have furnished proof that he no longer breathed. (247–48)

Interestingly, Faulkner links the nonlinear effects for which the lawyer cannot account with the moral virtues—love and honor—that Charles Bon will quixotically display. Deterministic thinking thus appears not only incapable of correctly assessing the world, but even morally deficient. To treat human beings as predictable components in a well-oiled machine reduces their humanity.

Sutpen's story, detailing the consequences of his design, is set in the framework of Faulkner's own design, one that necessarily involves a certain amount of deterministic thinking. It hardly takes the destabilizing assumptions of post-structuralist linguistics to make us aware that a writer's determinations, or more aptly intentions, cannot be realized. Even so, certain texts attempt, although futilely, to control our responses. *Absalom, Absalom!*, however, intentionally plays against its own intentionality. With its multiperspectival, nonlinear structure, it resists overdetermination or, at least, relishes the unpredictable response. In an oft-quoted response to an interviewer's question about the "truth" of the novel, Faulkner stated: "[T]he truth, I would like to think, comes out, that when the reader has read all these thirteen ways of looking at the blackbird, the reader has his own fourteenth image of that blackbird which I would like to think is the truth."[12] The text thus encourages, in Roland Barthes's terms, a "writerly" as opposed to a "readerly" reading, encompassing within its own structure an acknowledgment of the ongoing interpretive process that undermines a predictable truth about its meaning.[13] Yet, out of the plurality of meanings mobilized by its writerly nature emerges the structure of the strange attractor, bounding the text's plural in a zone of meaning.

The Attractor Structure of *Absalom, Absalom!*

I have so far spoken of the black blood as if it is a given. But it is not a given at all, at least not within the narrative proper—a particularly suspect designation with regard to this text. The only place where Bon's black blood appears as a "certainty" is in the "Chronology" affixed to the end of the text. Indeed, the "Chronology" and "Genealogy" serve as supplements (in the most saturated sense) to the narrative proper, for they not only restate the information provided there, but also provide information that appears to be unavailable to the internal narrators.[14] Although presenting the skeleton "facts" around which the several narrators weave their versions of the fall of the house of Sutpen, they also call attention to a lack of clear cause and effect and of origin of those facts. They undermine their own explanative function. The Chronology is akin to historical annals—those lists of facts that "represent historical reality as if real events did not display the form of a story."[15] In essence, they are events unanchored in plot. The plotlessness of the "Chronology" and "Genealogy" reinforces the text's implicit message that the observer must discern the causal pattern.

Nowhere in the narrative does an episode occur wherein an internal narrator presents a narratee with the "fact" of Bon's black blood—not at any of the temporal sites at which someone might be expected to do so. Whether we surmise that such a scene might have taken place is beside the point. Faulkner simply gives us an italicized passage wherein Quentin and Shreve have so fully engaged themselves in Henry and Bon's story that they seemingly become the very characters of whom they speak. In their version, Thomas Sutpen tells Henry that, after Bon's birth, he *"found out that his mother was part negro"* (283). But what are we to make of a "fact" brought forward during some apparently intuitive leap by Quentin and Shreve into the lives of their now-dead counterparts? An earlier passage in the novel suggests that italics signify "the long silence of notpeople, in notlanguage" (5), a nullification within discourse of discourse itself. The narrative thus elides an authoritative presentation of information. As Gerald Langford demonstrates in his comparison of the manuscript version with the public book, the elision appears to have been a deliberate move on Faulkner's part. In the manuscript version, Mr. Compson actually knows about Bon's black blood, and he transmits this information to Quentin.[16] The linear sequence of transmission is thus made clear—Thomas Sutpen to General Compson, General Compson to Mr. Compson, Mr. Compson to Quentin, and Quentin to

Shreve. By refusing to provide an explanation for Quentin's acquisition of this crucial piece of knowledge, Faulkner draws attention to its uncertain origin.

The text consistently emphasizes the indeterminable origins of the facts of Sutpen's life. Sutpen's own beginnings cannot be fixed, no more than Tristram Shandy's can; the initial conditions that produced him are unclear. According to what might be called a communal myth, Sutpen arrives in Jefferson "*out of nowhere and without warning*" (5). As Rosa Coldfield tells her tale, Quentin imagines he can see Sutpen and his slaves "drag house and formal gardens violently out of the soundless Nothing and clap them down like cards upon a table beneath the uppalm immobile and pontific, creating the Sutpen's Hundred, the *Be Sutpen's Hundred* like the oldentime *Be Light*" (4). In the beginning, in the void, Sutpen, like God proclaiming "fiat lux," always already is. Attempting to explain himself to General Compson, he cannot account for his presence in the Virginia Tidewater: "So he knew neither where he had come from nor where he was nor why. He was just there" (184).

To the internal narrators, Sutpen presents an impenetrable enigma. They know him mainly through various texts—carvings on headstones, entries in family Bibles, letters, and tales passed down through several generations. Each narrative reworking of Sutpen's story is predicated on a previous reworking. Even Rosa Coldfield and General Compson (those who actually knew the man) depend on others' stories (Ellen's, Sutpen's) to come up with their version of his life. Sutpen's own story comes to us through a layering of several intervening voices: Quentin tells Shreve the story as he heard it from Mr. Compson, who heard it from General Compson, who heard it from Sutpen himself. The narrators look at Sutpen from the outside, and they can only surmise what goes on in his mind, regularly qualifying any assertion about him with words such as "perhaps" and "doubtless." They often describe him in terms of an absence or a lack. He is the "nothusband" for whom (perhaps) Miss Rosa has worn black for forty-three years, "*a walking shadow*" who "*was not articulated in this world*" (139). In his attempt to explain the peculiar relationship between Miss Rosa and Sutpen, Mr. Compson provides a suggestive metaphor. He thinks that Sutpen's face must have seemed to Rosa "like the mask in Greek tragedy interchangeable not only from scene to scene but from actor to actor and behind which the events and occasions took place without chronology or sequence" (49). The passage serves as a self-reflexive comment on Faulkner's technique of emphasizing Sutpen's lack of ontological fixity.

Faulkner describes Charles Bon, as he does Sutpen, as lacking origin and corporeality. He suddenly impinges upon the town consciousness, "a personage who in the remote Mississippi of that time must have appeared almost phoenix-like, fullsprung from no childhood, born of no woman and impervious to time and, vanished, leaving no bones nor dust anywhere." (58). He is a nullity, "shadowy: a myth, a phantom: something which they engendered and created whole themselves; some effluvium of Sutpen blood and character, as though as a man he did not exist at all" (82). Rosa falls in love with Bon's "*pictured face*," which does "*not even need a skull behind it*" (118), but she never sees the actual man, even when he lies dead in Judith's room. When she helps to carry his coffin down the stairs, she tries "*to take the full weight of the coffin*" to prove to herself that he had indeed existed, but she cannot tell (122); she is left feeling that they "*had buried nothing*" beneath "*that mound vanishing slowly back into the earth*" (127). Rosa sums up Bon with a presciently post-structuralist paradox: "*he was absent, and he was; he returned, and he was not*" (123).

It is tempting to make this motif of absence and lack exactly that—a gaping hole at the center of the text around which the various narrative explanations endlessly spiral. But we would do better to envision Faulkner's structure in terms of a strange attractor, that multidimensional emplotment of a certain class of nonlinear systems. There *are* real "facts"—an actual "history"—around which the narrative explanations twine, but they are unreachable in themselves. Nevertheless, the trajectories of the narratives fall onto an attractor generated by those unreachable facts.

The attractor begins to take shape when we examine the multiperspectival narratative structure of *Absalom, Absalom!*. In the text, Sutpen's story ostensibly is puzzled over by four internal narrators—Rosa Coldfield, Mr. Compson, Quentin Compson, and Shreve McCannon—whose narratives are themselves presided over by the external narrator, seemingly there to locate us in place and time and provide stage directions as the characters engage in their incessant discourse. I say "ostensibly" because no clear-cut boundaries exist between the narratives of Quentin and Shreve and because other narratives are embedded within the main internal narratives—Sutpen's story to General Compson, General Compson's story to Mr. Compson, Bon's letter to Judith, and so forth. In addition, the boundary between the external and the internal narratives is itself fuzzy, with Sutpen's story at times seemingly coming from the external narrator yet tenuously focalized through an internal narrator. We often cannot know where one voice ends and another begins.

Out of this cacophony of voices, however, the plot of the text emerges, just as the attractor pattern emerges when the evolving behavior of a chaotic dynamical system is plotted in state space. We can compare each of the stories that is told about Sutpen to a trajectory upon the strange attractor. Although the strange attractor comprises one infinitely long, never-repeating trajectory, different sets of initial conditions give rise to trajectories that visit different sections of the attractor. With regard to time, proximity, and connection to the principal players in the events at Sutpen's Hundred, each of the internal narrators begins from a different set of initial conditions. Rosa Coldfield, on the one hand, is seemingly closest in all respects: she actually knew all the principal players except Bon and lived for a time at Sutpen's Hundred. Shreve McCannon, on the other hand, is furthest away: he was born after all the players except Henry, Clytie, and Rosa had died; he hears of the events at second-, third-, and fourth-hand; and he has never even been to the South, let alone to the cursed parcel of land where Sutpen's tragedy played itself out.

Beginning from these sets of initial conditions, the external and internal narratives trace trajectories through the state space of the text. As Sutpen's story is (re)iterated, we begin to see that these trajectories fall onto an attractor. All of the trajectories are attracted to, but never pass directly through, what seems to be the crucial event—the revelation of Bon's black blood. That event serves as the unstable attracting point, concurrently attracting and repulsing the internal narratives. Henry's shooting of Bon, the apparent fatal consequence of this revelation, is another unstable attracting point, for Faulkner gives us no definitive knowledge about it. He presents the episode as an imaginative recreation by Quentin, interrupting an otherwise seamless narrative by Mr. Compson, which elides the actual encounter. In effect, each internal narrative lies on the attractor that constitutes the plot of *Absalom, Absalom!*.

Because each narrator begins from a different set of initial conditions, the story he or she tells falls onto the attractor at a different point. The further away from the time of the events that the narratives occur, the more those narratives not only incorporate the information of preceding iterations but also replace it with new information. Initial conditions are themselves lost; we have no precise way of knowing how any of the narratives got the way they are—which "rag-tag and bob-ends of old tales and talking" (243) went into the reconstructions of the various narrators. As with a chaotic system, at certain times, trajectories almost converge; at other times, they diverge widely. At certain times, trajectories come close to the attracting point; at other times, they veer away. Some trajectories such as Rosa Coldfield's rehashing of Sutpen's insulting behavior visit certain regions more frequently. (Interestingly, the insult itself is never actually stated). Each reiteration fills in more information but diverges from what has come before so that the narrative never exactly retraces its own footsteps.

In chaotic systems, once we start from a set of initial conditions, we have no way of knowing when and thus where the trajectory will come close to the attracting point or veer away from it. If we regard the veiled revelation of Bon's black blood as the absent event that all the trajectories attempt to reach, the joint narrative trajectory of Quentin and Shreve comes closest to this unstable attracting point. The temptation has been to view their joint narrative as most accurately explaining what happened at Sutpen's Hundred. Certainly, it makes fall into place pieces of the puzzle that have previously not seemed to fit: Sutpen's repudiation of Bon; Henry's subsequent shooting of Bon; Judith's quasi-adoption of Charles Etienne; and Charles Etienne's Sutpen-like "furious and indomitable desperation" (164). There has also been a temptation to view their narrative as most accurately explaining what happened because it can draw upon preceding iterations of Sutpen's story. But should we do so? No and yes. To a certain extent, the text deliberately works against our regarding the joint narrative as some sort of Sherlockian solution pieced together once all the evidence has been gathered. Again, the actual transmission of the vital piece of information never occurs between any narrator and narratee. Furthermore, the logic that holds the joint narrative together is put on equal footing with other sorts of narrative "glue." The story that Quentin and Shreve tell has no more basis in fact than the stories told by the other narrators: "the two of them, whether they knew it or not, in the cold room, (it was quite cold now) dedicated to that best of ratiocination which after all was a good deal like Sutpen's morality and Miss Coldfield's demonizing." (225). Again and again, the text calls attention to the fact that the joint narrative is a reconstruction of other reconstructions, themselves based on suspect knowledge. The temporal and geographical distancing of the narrating instance—Harvard, 1910—from the actual events further reinforces the problematic nature of the knowledge to which Quentin and Shreve laid claim.

This distancing receives additional emphasis with Shreve's assumption of the role of sole narrator. Shreve does not simply retell in a distinctive style something Quentin has told him previously. Granted, Shreve *has* garnered information about the

South from Quentin, as the following passage makes clear: " 'How was it?' Shreve said. 'You told me; how was it?' " (152). But Shreve tells Quentin things that Quentin does not already know. Significantly, he alternates between surmise and assertion, between not knowing and knowing. He does not attempt to guess what goes on in the mind of Judith Sutpen: "And the girl, the sister, the virgin—Jesus, who to know what she saw that afternoon when they rode up the drive" (256). But he does claim to know Charles Bon's innermost thoughts. He asserts that it was Henry, not Bon, who was wounded at Shiloh, flatly contradicting Mr. Compson's opposing information and actually calling any other narrator's access to truth into question: "Because your old man was wrong here, too! He said it was Bon who was wounded, but it wasn't. Because who told him? Who told Sutpen, or your grandfather either, which of them it was who was hit?" (275). Yet the very doubt Shreve casts on the other narrator's lack of direct knowledge inevitably rebounds on him. We are explicitly told that Shreve's invents his "facts": "four of them who sat in that drawing room of baroque and fusty magnificence which Shreve had invented and which was probably true enough." (268). And when he takes over the narrative from Quentin, he demands, "Let me play a while now" (224). Clearly, it is not knowledge per se that gives Shreve's narrative an explanatory power greater than that of the other narratives.

At the same time, however, we are told that Shreve's invention is "probably true enough." Thus Shreve's lack of direct knowledge is complemented by insights that enable the pieces of the puzzle to fall into place. His accession to such insight makes sense in light of a chaos-theory model. The joint narrative does not result from his piecing together of partial information from other iterations of Sutpen's story. Shreve's insights instead come from an intuitive leap—a leap onto the attractor structure that can be discerned through the previous iterations.

Shreve's leap is analogous to that made by the dynamicist who watches as a trajectory evolves in state space and discerns the emergent pattern. A strange attractor exists only insofar as it is predicted by the dynamicist viewing the simulation.[17] A feedback loop exists between the dynamicist and the simulation whereby the dyanmicist manipulates the parameters in order to find the most accurate interpretation of the system being modeled.[18] By taking into account the feedback loop between the modeler and the model, chaos theory foregrounds the way in which the emerging pattern of the attractor is not simply observed but also influenced by the dynamicist.

Just so does *Absalom, Absalom!* take into account the way in which the reader of Sutpen's story influences the narrative and is influenced by it, functioning as both narratee and narrator. Shreve performs a writerly reading of the other narratives through his intuitive leap onto the attractor upon which their narrative trajectories fall. But this process is more than simply a conventional reader-response approach of filling in gaps; Shreve's iteration simultaneously adds to the pattern and changes it. Although the original information is irrecoverable, the information that remains is similar to what remains as a trajectory evolves in a dynamical system, which Thomas Weissert explains thus: "Whereas the part of meaning that derives from the external relations of the dynamical signifiers slips away as the exponential separation of orbit pairs, there remains a residual meaning sustained by the purely internal relations of the mathematical signification system."[19] The "internal relations" of the Sutpen narrative are not static, but dynamic, entangled in a feedback loop with Shreve.

Our own insights as to the truth about Sutpen—always approximate, never fully revealed—emerge from the ensemble of narrative iterations that make up the novel. We do not privilege that final narrative trajectory jointly put forward by Quentin and Shreve, but instead take into account all the previous trajectories, discerning the attractor upon which they fall and influencing it in our turn. This emergent ensemble is Faulkner's "fourteenth image of the blackbird." To some extent, we, as well as Shreve, leap onto the attractor, our commentary on the text constituting yet another trajectory that presumably adds to the pattern and changes it, ensuring that we can never recover the initial conditions of an innocent reading. Shreve thus serves within the text to model the reading experience that the text invites us to have. He is both the model reader and the model for the reader.

Again, the attractor structure of *Absalom, Absalom!* arises from a suppressed piece of information in the text—the repudiation of the black Other by the white master. Structure and theme are integrally entangled, the attractor structure serving as a means for reinforcing the theme of the novel. The truth that the Canadian Shreve approximates is one that the Southern narrators cannot directly acknowledge—thus the fact that the narrator furthest from the events comes closest to their meaning. This truth concurrently attracts and repulses because actually to arrive at it is to undermine the (patri)linear master narrative of the rise of the "civilized" South. Sutpen must deny his own issue in order to preserve the purity of his design.

That supposed purity was, however, never there in the first place; Sutpen's dynastic ambitions, like those of his fellow Southerners, can only be achieved at the expense of black blood.[20] Sutpen is not the quintessential self-made man. Slave labor enables him to wrest his mansion out of the "soundless Nothing" and supply it with the accoutrements of Southern gentility. Sutpen's wife and daughter, like other Southern wives and daughters, can keep their feet warm during a carriage ride on a winter's day only because there is "an extra nigger on the box with the coachman to stop every few miles and build a fire and re-heat the bricks on which Ellen's and Judith's feet rested" (81–82). The plantation belles' virginity stays intact because of the rape of their black sisters, as Mr. Compson points out:

> [T]he other sex is separated into three sharp divisions, separated (two of them) by a chasm which could be crossed but one time and in but one direction—ladies, women, females—the virgins whom gentlemen someday married, the courtesans to whom they went while on sabbaticals to the cities, the slave girls and

women upon whom that first caste rested and to whom in certain cases it doubtless owed the very fact of its virginity. (87)

The myth of Southern gentility can be preserved only because the fact of Southern brutality is suppressed.

Through the Sutpen story, the text demonstrates the fractal property of similarity across scale, consistent with a dynamical systems model. The local event of Sutpen's denial of Bon is replicated globally in the Southerners' denial that the black Other whom they exploit and brutalize to erect their stately mansions is, in fact, kindred: "all boy flesh that walked and breathed stemming from that one ambiguous eluded dark fatherhead and so brothered perennial and ubiquitous everywhere under the sun—" (240). The poignant exchange between Henry and Bon after Sutpen's revelation reinforces this theme:

   —You are my brother.
   —No I'm not. I'm the nigger that's going to sleep with your sister. (286)

Race cancels out common humanity. Sutpen's progeny, with their legacy of fratricide and potential miscegenation and incest, enact in a local framework the curse that slavery brought to the South.

Like the other writers that I have examined, Faulkner was not privy to chaos theory—although it can be argued that, like them, he was capable of articulating the incipient elements in the cultural matrix.[21] The attractor structure of *Absalom, Absalom!* points toward his dissatisfaction with the overdetermination of linear narrative and its spurious claims of revealed knowledge. Faulkner's gift lies in his ability not only to apprehend the dynamics of narrative, with its resistance to totalizing perspectives, but also to foreground this dynamics to encourage our active engagement in the process. Perhaps more importantly, however, the attractor structure brings to the fore the very difficulty of the historical problem with which Faulkner's text deals—the terrible consequences of slavery. As a Southern writer,

Faulkner had to address that issue, but he was immersed in a culture that itself could not come to terms with it. When Faulkner shows his characters' difficulty with getting at the truth about Sutpen, he mirrors his own difficulty as a Southern writer concurrently drawn to and repulsed by the tragic story he tells. Quentin Compson's ambivalent final words about the South may reflect Faulkner's own feelings: "*I dont hate it! I dont hate it*" (303). Applying the attractor model to Faulkner's text enables us to understand or, perhaps more accurately, approximate the conditions from which it arises.

The epigraph at the beginning of this chapter may provide the most fitting analogy of the workings of Faulkner's narrative—and perhaps narrative in general. When we throw a pebble into a pond, it becomes as inaccessible as the attracting point in a chaotic system. All that we can see is the pattern made by its ripples, a continuously evolving configuration. But, as Faulkner envisions, "*that pebble's watery echo*" moves across the water "*at the original ripple-space*" (210), a testimony to the irrecoverable attracting point that determines the pattern in the first place.

The attractor model, implicit in Faulkner's text, gives us a way of assessing our own critical procedures. When we evaluate a text, we get on the attractor, our own narrative trajectory approaching an unstable truth and then veering away from it. If we had infinite time, so that we could turn (state) space into a Borgesian library filled with commentary, a truth might emerge—about Sutpen, this text, and texts in general. As it is—as Faulkner shows us—we can only make our way along the attractor, knowing that we will never arrive at the attracting point and that our own trajectory is just one segment in the ever-evolving whole.

Postscript

The noise is the source from which all stories emerge.
—Italo Calvino, "How I Wrote One of My Books"

Disorder is never anything but a different order than we expect.
—Jean Guitton in *Chaos and Determinism*

It might seem that what I have said about chaotic narratives could be applied to all narratives—a universal theory of bounded randomness. However, we must bear in mind that strange attractors are not a property of all dynamical systems, not even of all chaotic systems, but only of a particular type of chaotic system. If we take the strange-attractor structure as an all-purpose model for narrative dynamics, a new formalism that can be applied to all narratives, we risk committing the same errors that formalist, structuralist, and certain types of post-structuralist theories have committed. The one-size-fits-all approach erases the social, cultural, and historical circumstances out of which narratives arise. Just as strange attractors are a property of only *some* nonlinear systems, in narrative dynamics, they are a property of only *some* kinds of narratives—narratives whose meaning can only be conveyed through a species of disorderly order.

Writers make deliberate choices about the structure their narratives will assume, and chaotic narratives, as I have defined them, are thus outgrowths of the particular circumstances under which their writers operate, the form integrally related to the thematic content. The strange attractors that I have located in the four texts differ, just as they differ in physical systems in which they are found—the strange attractor generated by the dynamics of the dripping faucet unlike that generated by the dynamics of weather patterns, for example. Yet for each text in this study, the strange-attractor structure arises from powerful thematic content that simultaneously attracts and repels the writers. I would suggest that the bounded randomness of deterministic chaos accrues to a particular subset of narratives, such as the ones that I have examined.

Although I do not put forward the strange-attractor model as a global theory about textual dynamics, we need to consider whether all narratives can be evaluated by means of chaos theory. By drawing on these insights the theory makes available to us, we can begin to explore the analogy between dynamical systems and narratives, and, as I suggest in chapter 1, such a line of inquiry is a promising one. By considering a narrative as a dynamical system, we can mediate between notions of narrative as spatial product and temporal process. Narratives can be regarded as fixed patterns whose meaning continuously evolves. As chaos theory demonstrates, simulations of dynamical systems result from a feedback loop between system and dynamicist. Interpretations of narratives similarly result from a feedback loop between text and reader. Although the quality of bounded randomness is a structural element of the four texts I have examined, we can also apply it to the process of meaning-making in general. Narratives—even linear ones—are neither fixed in meaning nor radically indeterminable. They can instead be considered as globally deterministic systems allowing an infinite play of variations at the local level.

In the sciences, chaos theory has enabled us to discern the deterministic chaos inherent in certain dynamical systems and to apprehend important truths about the process of meaning-making in which all scientists engage. In narrative studies, at the local level, chaos theory can help us to discover the interaction between form and meaning in chaotic texts. At the global level, it can help us to apprehend that all our narratives emerge from the noise that is their source.

NOTES

PREFACE

1. Tom Stoppard, *Arcadia* (London: Faber and Faber, 1993) 79.
2. Ivar Ekeland discusses the plight of Johannes Kepler, who attempted to chart the trajectories of planets in the early seventeenth century. Despite having sound theories upon which to draw, "He nevertheless had to perform monstrous computations over a number of years." Ekeland explains that, even with the digital computers of today, "There are still a great many computations that cannot be performed now or in any foreseeable future." See Ivar Ekeland, *Mathematics and the Unexpected* (Chicago: University of Chicago Press, 1988) 5, 31; trans. of *Le Calcul, l'imprevu: Les figures de temps du Kepler à Thom* (Éditions du Seuil, 1984).
3. The phrase is James Gleick's. His popular science book *Chaos: Making a New Science* (New York: Viking, 1987) brought chaos theory into the public imagination.
4. I discuss these works in chapter 1.

CHAPTER I CHAOS THEORY AND THE DYNAMICS OF NARRATIVE

1. James P. Crutchfield, J. Doyne Farmer, Norman H. Packard, and Robert S. Shaw, "Chaos," *Scientific American* December 1986: 49. These four were members of the Dynamical Systems Collective of the University of California at Santa Cruz. "Chaos" was one of the first, if not the first, popular texts on the subject of chaos theory.
2. As Crutchfield et al. point out, "The larger framework that chaos emerges from is the so-called theory of dynamical systems" (49).
3. N. Katherine Hayles discusses the concurrent scientific imprecision and cultural resonance of the terms "chaos theory" and "science of chaos" in *Chaos Bound: Orderly Disorder in Contemporary*

Literature and Science (Ithaca: Cornell University Press, 1990). She points out that "[t]he older resonances" of the term chaos "linger on, creating an aura of mystery and excitement that even the more conservative investigators into dynamical systems methods find hard to resist" (8–9).
4. Ibid., *Chaos Bound*, 6.
5. Julian C. R. Hunt, Foreword, *Chaos and Determinism: Turbulence as a Paradigm for Complex Systems Converging toward Final States*, by Alexandre Favre, Henri Guitton, Jean Guitton, André Lichnerowicz, and Etienne Wolff, trans. Bertram Eugene Schwarzbach (Baltimore: Johns Hopkins University Press, 1995); trans. of *De la causalité à la finalité. A propos de la turbulence* (Paris: Éditions Maloine, 1988), xvii.
6. Stephen H. Kellert, *In the Wake of Chaos: Unpredictable Order in Dynamical Systems* (Chicago: University of Chicago Press, 1993), 145.
7. Peter Covenay and Roger Highfield, *The Arrow of Time: A Voyage through Science to Solve Time's Greatest Mystery* (New York: Fawcett Columbine, 1990), 64. Ivar Ekeland claims, "The most perfect mathematical expression of determinism is the differential equation." See *Mathematics and the Unexpected* (Chicago: University of Chicago Press, 1988), 20; trans. of *Le Calcul, l'imprévu, Les figures de temps du Kepler à Thom* (Éditions du Seuil, 1984).
8. Crutchfield et al., 49.
9. Kellert, 134–35. See the entire chapter "Beyond the Clockwork Hegemony" (119–58) for an enlightening discussion of the neglect of nonperiodic systems in classical physics.
10. James Gleick discusses Lorenz's case in his chapter "The Butterfly Effect," in *Chaos: Making a New Science* (Viking: New York, 1987), 9–31. See also Ian Stewart's chapter "The Weather Factory," in *Does God Play Dice: The Mathematics of Chaos* (New York: Basil Blackwell, 1989), 127–44.
11. Ilya Prigogine and Isabelle Stengers, *Order out of Chaos: Man's New Dialogue with Nature* (Toronto: Bantam Books, 1984), 75.
12. Favre et al., 146.
13. Hunt, xvii.
14. Ekeland, 18.
15. Sir Isaac Newton, *Sir Isaac Newton's Mathematical Principles of Natural Philosophy and His System of the World*, trans. Andrew Motte, revised trans. Florian Cajori (1729; Berkeley: University of California Press, 1946), 6.

16. Prigogine and Stengers, 17.
17. Michel Serres with Bruno Latour, *Conversations on Science, Culture, and Time*, trans. Roxanne Lapidus (Ann Arbor: University of Michigan Press, 1995), 60–61 trans. of *Éclaircissements* (Éditions François Bourin, 1990).
18. G. J. Whitrow, *Time in History: The Evolution of Our General Awareness of Time and Temporal Perspective* (Oxford: Oxford University Press, 1988), 127. John Casti also comments upon the connection between the Newtonian conception of time and the clock: "When we use the Newtonian time axis to represent a set of observed real-world events, we try to produce somehow a 'clock' whose time moments (the vertices) can be put into one-to-one correspondence with the set of events." See John L. Casti, *Complexification: Explaining a Paradoxical World through the Science of Surprise* (New York: HarperCollins, 1994), 200.
19. Prigogine and Stengers, 46.
20. Evelyn Fox Keller, *Reflections on Gender and Science* (New Haven: Yale University Press, 1985), 69.
21. Prigogine and Stengers, 50.
22. Alexander Pope, *Poetry and Prose of Alexander Pope*, ed. Aubrey Williams (Boston: Houghton Mifflin, 1969), 130.
23. Stewart, 17.
24. Favre, 21.
25. Kellert, 144–45.
26. Edward N. Lorenz, "Deterministic Nonperiodic Flow," *Journal of the Atmospheric Sciences* 20 (1963): 130; rpt. in *Chaos*, ed. Hao Bai-Lin (1984; Singapore: World Scientific, 1985), 282.
27. Ibid., 141 (293).
28. Stewart, 134. Stewart notes that the venue in which Lorenz published probably contributed to the ignoring of Lorenz's paper: "The topologists, whose necks would doubtless have prickled like mine that they come across Lorenz's seminal opus, were not in the habit of perusing the pages of the *Journal of Atmospheric Sciences*" (134).
29. Robert Shaw, *The Dripping Faucet as a Model Chaotic System* (Santa Cruz: Aerial Press, 1984), 1–2.
30. For a detailed explanation of the dripping faucet's behavior, see Shaw, 14–16.
31. Gleick, 264.
32. Favre et al., explain that, for chaotic systems, "the behavior of fluids may not violate the laws of mechanics and physics.

However, these necessary general conditions are not sufficient for a complete determination of a fluid flow; the particular circumstances of each flow must be taken into account" (44).
33. Determinism, as Ekeland points out, "can only be the property of reality as a whole of the total cosmos." But this reality is unavailable to us: "Global reality, the cosmos taken as a whole, from the most minute elementary particle to the expanding universe, is out of our reach. Science can only isolate subsystems for study, and set up experimental screens on which to project this inaccessible whole. Even if reality is deterministic, it may well happen that what we observe in this way is unpredictability and randomness" (62).
34. Crutchfield, 51.
35. Ibid., 49. According to Leon Glass and Michael Mackey, "although in principle it should be possible to predict future dynamics as a function of time, this is in reality impossible since any error in specifying the initial conditions, no matter how small, leads to an erroneous prediction at some future time." See Leon Glass and Michael C. Mackey, *From Clocks to Chaos: The Rhythms of Life* (Princeton: Princeton University Press, 1988), 6–7.
36. Lorenz, 133 (285). Gleick describes this situation at length in his chapter on "The Butterfly Effect," in *Chaos* 11–31. With regard to the weather, Ekeland points out, "It is estimated that small perturbations are multiplied by 4 every week, or by 300 every month" (66). As I was writing an early draft of this chapter, I was gazing out the window at piles of snow, the result of a completely unexpected nor'easter. In an attempt to explain why the storm had not been predicted, the metropolitan daily called up a butterfly-effect scenario: "The biggest snowfall in four years was the result of a peculiar alignment of circumstances that no computer could model accurately. It had to do with a stalled storm, a pool of cool air in the upper atmosphere, the tepid waters of the Gulf Stream, and even last week's snow." See Anthony R. Wood, "How Weather Forecasters Got Snowed," *The Philadelphia Inquirer* 26, January 2000: A12.
37. As Covenay and Highfield note, the emergence of chaotic structures "in the simplest of situations . . . demolishes the centuries-old myth of predictability and time-symmetric determinism, and with it any idea of a clockwork universe" (37).

38. J. T. Fraser, "From Chaos to Conflict." *Time, Order, Chaos: The Study of Time IX*, ed. J. T. Fraser, Marlene Soulsby, and Alexander Argyros (Madison, CT: International Universities Press, 1998), 12.
39. Certainly, the Second Law of Thermodynamics and the Special Theory of Relativity challenged Newtonian notions of time. The Second Law presupposes an irreversible "arrow of time," contrary to the time-symmetric quality of Newtonianism. According to Whitrow, whereas Newton's concept of time is independent of the universe, Einstein's concept that time is relative has prevailed: "[T]he condition that each event has only one time associated with it no longer holds. Instead, its time depends on the observer" (173). Nevertheless, as Alexander Argyros points out, even relativistic time is conceived of in terms of a fixed, external phenomenon: "[T]ime has been typically assumed to be simply 'out there,' a fundamental component of reality. Even the most unsettling revolutions in the way we think about time, the theories of special and general relativity, share this basic assumption." See *A Blessed Rage for Order: Deconstruction, Evolution, and Chaos* (Ann Arbor: University of Michigan Press, 1991), 130.
40. Kellert offers the following description of the observer's new role: "Far from creating a space for the reappearance of qualitative properties in the sense of subjective, sensuous experiences, chaos theory strives to apply mathematical techniques to phenomena like turbulence that were once a repository for Romantic notions of sublime Nature resisting the onslaught of human rationality" (115).
41. Edgar Morin, "The Fourth Vision: On the Place of the Observer," trans. Pierre Saint-Amand, *Disorder and Order: Proceedings of the Stanford International Symposium (September 14–16, 1981)*, ed. Paisley Livingston (Saratoga, CA: Anma Libri, 1984), 103.
42. Ibid., 106.
43. Thomas P. Weissert, *The Genesis of Simulation in Dynamics: Pursuing the Fermi-Pasta-Ulam Problem* (New York: Springer, 1997), viii.
44. Prior to the computer age, we could, of course, map the behavior of periodic systems. However, in order to map the behavior of a nonperiodic system, we needed to perform so many iterations of the equations that it was unfeasible to do so.

45. The phrase comes from Crutchfield et al., 50.
46. Thomas Weissert, "Dynamical Discourse Theory," *Time and Society* 4 (1995): 118.
47. Shaw, 17.
48. Crutchfield et al., 46.
49. Michael Berry, "Chaology: The Emerging Science of Unpredictability," *Royal Institution Proceedings* 61 (1989): 202.
50. Lorenz, 137 (289).
51. I am indebted to Weissert for the phrase "bounded randomness." He uses it "to indicate that we have elements of both determinations and randomness. Because the trajectory on the attractor resides within a three-dimensional Cartesian space, the signifying point can trace out a path within the bounded region that never crosses itself, never repeats itself exactly, and never comes to rest on any one single point" ("Dynamical Discourse," 122).
52. Crutchfield et al., 53. Hayles notes, "Attractors operate irreversibly because their operation changes *what we can know about them*, not merely what we do know" (Emphasis in the original; *Chaos Bound*, 159).
53. Stewart elaborates on the connection between chaos and fractals: "During the 1970's, when both were in their infancy, chaos and fractals appeared unrelated. But they are mathematical cousins. Both grapple with the structure of irregularity. In both, geometric imagination is paramount. But in chaos, the geometry is subservient to the dynamics, whereas in fractals the geometry dominates. Fractals present us with a new language in which to describe the shape of chaos" (222).
54. Ibid., 223.
55. Ekeland 46, 46–47.
56. The University of Texas at Austin hosts the Center for Nonlinear Dynamics in the physical sciences while McGill University hosts the Centre for Nonlinear Dynamics in Physiology and Medicine. The Society for Chaos Theory in Psychology and the Life Sciences features an annual conference of scholars throughout the disciplines.
57. Crutchfield et al., 57. Interestingly, Prigogine/Stengers and Crutchfield et al. represent two different branches of chaos theory—what Hayles calls, respectively, the "order-out-of-chaos" branch, and the "strange-attractor" branch (*Chaos Bound* 9–10). Yet each branch points to humanistic applications of

chaos theory, and these applications depend on the insights that each branch makes available to us.
58. Tom Stoppard, *Arcadia* (London: Faber and Faber, 1993); Darren Aronofsky, writer and director, *Pi* (Artisan Entertainment, 1998); Michael Crichton, *Jurassic Park* (New York: Knopf, 1990).
59. For the link between post-structuralism and chaos theory in Hayles, see her chapter "Chaos and Poststructuralism," *Chaos Bound*, 175–208, and in Weissert, see "Dynamical Discourse Theory," *Time and Society* 4 (1995): 111–33. Although Hayles and Weissert argue for, respectively, an isomorphic and an isotropic connection between the methods of Derridean deconstruction and modern dynamics, each points out that, rather than repudiating order, as deconstruction seems to do, chaos theory puts forward new possibilities for it. Argyros draws upon the tenets of chaos theory to mount a sustained attack against the valorization of randomness and relativity that he regards as common to post-structuralist thought. For Hayles's discussion of the cultural implications of chaos theory, see her chapters "The Politics of Chaos: Local Knowledge versus Global Theory" and "*Conclusion:* Chaos and Culture: Postmodernism(s) and the Denaturing of Experience," *Chaos Bound*, 209–35, 265–95.
60. Colin Martindale, "Chaos Theory, Strange Attractors, and the Laws of Literary History," *Empirical Studies of Literature: Proceedings of the Second IGEL-Conference, Amsterdam 1989*, ed. Elrud Ibsch, Dick Schram, and Gerard Steen (Amsterdam: Rodopi, 1991): 381–85.
61. Notable chaos-theory readings include N. Katherine Hayles, "Chaos as Dialectic: Stanislaw Lem and the Space of Writing" and "Fracturing Forms: Recuperation and Simulation in *The Golden Notebook*," *Chaos Bound*, 115–40, 236–64; Istvan Csicsery-Ronay, Jr., "Modeling the Chaosphere: Stanislaw Lem's Alien Communications," *Chaos and Order: Complex Dynamics in Literature and Science* ed. N. Katherine Hayles (Chicago: University of Chicago Press, 1991), 244–62; Thomas P. Weissert, "Representation and Bifurcation: Borges's Garden of Chaos Dynamics," *Chaos and Order*, 223–43; Paul Harris, "Fractal Faulkner: Scaling Time in *Go Down, Moses*," *Poetics Today* 14 (1993): 625–51; Richard Nemesvari, "Strange Attractors on the Yorkshire Moors: Chaos Theory

and *Wuthering Heights*," *The Victorian Newsletter* 92 (1997): 15–21; and Timothy Jackson Rice, *Joyce, Chaos, and Complexity* (Urbana: University of Illinois Press, 1997). Gordon E. Slethaug examines contemporary American fiction through the chaos-theory lens in *Beautiful Chaos: Chaos Theory and Metachaotics in Recent American Fiction* (Albany: State University of New York Press, 2000). Philip Kuberski discusses chaos theory in his *Chaosmos*, regarding it as part of a larger movement in a "postmodern" science that is reconnecting humanity to nature; see *Chaosmos: Literature, Science, and Theory*, The Margins of Literature (Albany: State University of New York Press, 1994). Other full studies that draw on chaos theory include Harriet Hawkins's *Strange Attractors: Literature, Culture, and Chaos Theory* (New York. Harvester Wheatsheaf-Prentice Hall, 1995); Hans C. Werner's *Literary Texts as Nonlinear Patterns: A Chaotics Reading of* Rainforest, Transparent Things, Travesty, *and* Tristram Shandy, Gothenberg Studies in English 75 (Göteborg, Sweden: Acta Universitatis Gothoburgensis, 1999); and Emily Zants's *Chaos Theory, Complexity, Cinema, and the Evolution of the French Novel*, Studies in French Literature 25 (Lewison, NY: The Edwin Mellen Press, 1996). For a collection of chaos-theory readings of eighteenth-century texts, see Theodore E. D. Braun and John McCarthy, eds., *Disrupted Patterns: On Chaos and Order in the Enlightenment*, Internationale Forschungen zur Allgemeinen und Vergleichenden Literaturwissenschaft 43 (Amsterdam-Atlanta, GA: Rodopi, 2000).

62. Paul R. Gross and Norman Levitt, *Higher Superstition: The Academic Left and Its Quarrels with Science* (Baltimore: Johns Hopkins University Press, 1994), 5, 6. The text is a self-proclaimed "jeremiad"; its main goal apparently is to attack postmodern theory in general, including what the authors regard as manifestations of it, such as feminism, social constructivism, and multiculturalism.

63. The original essay "Transgressing the Boundaries: Toward a Transformative Hermeneutics of Quantum Gravity" was published in *Social Text* 46/47 (1996): 217–52. It has since been reprinted as an appendix in Alan D. Sokal's and Jean Bricmont's *Fashionable Nonsense: Postmodern Intellectuals' Abuse of Science* (New York: Picador USA, 1998) 212–58; trans. of *Impostures Intellectuelles* (France: Éditions Odile Jacob, 1997). Sokal revealed the parody in "A Physicist Experiments with

Cultural Studies," *Lingua Franca* (May/June 1996): 62–64. He explained his reasons in "Transgressing the Boundaries: An Afterword," *Dissent* 43 (1996): 93–99. This essay is also reprinted in *Fashionable Nonsense*, 268–80.

64. Alan D. Sokal, "What the *Social Text* Affair Does and Does Not Prove," *A House Built on Sand: Exposing Postmodernist Myths about Science*, ed. Noretta Koertge (New York: Oxford University Press, 1998), 11.
65. Braun and McCarthy, Foreword, vi.
66. Sokal and Bricmont, 11. Countering this view, Stanley Aronowitz dryly comments upon the tendency of restricting discussion of science to scientists: "While everybody, including physicists and molecular biologists, is qualified to comment on politics and culture, nobody except qualified experts should comment on the natural sciences." See Stanley Aronowitz, "The Politics of the Science Wars," *Science Wars*, ed. Andrew Ross (Durham: Duke University Press, 1996), 203.
67. Rice notes that "literature, science, and the communities working within and affected by both disciplines inhabit an even larger relational 'field of activity': literature and science are fundamental constituents of *one* culture, the only one we have" (x). Part of the defensiveness of scientists such as Sokal and Bricmont may derive from a concern that to regard science as part of culture is to regard it as a cultural construction. Sokal and Bricmont, for example, decry "epistemic relativism": "the idea that modern science is nothing more than a 'myth,' a 'narration' or a 'social construction' among many others" (x). Gross and Levitt declare flatly that the contention that scientific knowledge is ideological is "wrong" (253).
68. Steven Johnson, "Strange Attraction," *Lingua Franca: The Review of Academic Life* 6: 3 (1996): 47.
69. Ibid., 50. Johnson refers here to the Santa Fe Institute, a scientific research center focused on complexity science.
70. Paul Ricoeur, *Time and Narrative*, vol. 1, trans. Kathleen McLaughlin and David Pellauer (Chicago: University of Chicago Press, 1984), 48; trans. of *Temps et Recit* (Paris: Éditions du Seuil, 1983).
71. Ibid., 53.
72. Gérard Genette's discussion of the triad "story," "narrating," and "narrative" in *Narrative Discourse Revisited* is another way of thinking about the structuration process pertaining to narrative, although Genette does not emphasize the dynamic

aspect of it. See *Narrative Discourse Revisited*, trans. Jane Lewin (Ithaca: Cornell University Press, 1988), 13–15; trans. of *Nouveau discours du récit* (Paris: Éditions de Seuil, 1983).
73. Thomas P. Weissert, "Dynamics and Narrative: The Time-Identity Conjugation," *Time, Order, Chaos*, 164. Focusing on the modeling process itself, Weissert remarks upon "the analogy between a text—considered as a complex trajectory space of narrative and signification, generated by an author's conscious and unconscious model of cultural relations—and the phase space of a dynamical system, generated from the dynamicist's model of the relations among the degrees of freedom of the referent physical system" ("Dynamical Discourse," 124). Harris discusses "the analogy between a processual literary form and a complex dynamical system" in terms of an interaction between the system and something external to it: "both are open systems, or systems that interact with their environments, in which internal complexity builds up over time, resulting in a commingling of the many possible forms that the system may realize" (642).
74. Fraser, 5.
75. Genette, *Narrative Discourse Revisited*, 15.
76. *Reading for the Plot: Design and Intention in Narrative* (1984; New York: Vintage, 1985), 25.
77. As is no doubt apparent, I am more than a little indebted to the taxonomy of narrative devices put forward by Gérard Genette in *Narrative Discourse: An Essay in Method*, trans. Jane E. Lewin (1980; Ithaca: Cornell University Press, 1983).
78. See Genette's definition of the iterative in *Narrative Discourse*, 116.
79. Casti, 29.
80. Roland Barthes, *S/Z: An Essay*, trans. Richard Miller (New York: Hill and Wang, 1974); trans. of *S/Z* (Paris: Éditions du Seuil, 1970), 5.
81. "Cultural Feminism versus Post-Structuralism: The Identity Crisis in Feminist Theory," *The Second Wave: A Reader in Feminist Theory*, ed. Linda Nicholson (New York: Routledge, 1997), 339.
82. Genette, *Narrative Discourse Revisited*, 8.
83. Susan Sniader Lanser, *Fictions of Authority: Women Writers and Narrative Voice* (Ithaca: Cornell University Press, 1992), 4.
84. Franco Moretti, *The Modern Epic: The World System from Goethe to García Márquez*, (London: Verso, 1996), 6.

CHAPTER 2   NARRATING AGAINST THE
CLOCKWORK HEGEMONY: *TRISTRAM
SHANDY*'S GAMES WITH TEMPORALITY

1. Laurence Sterne, *The Life and Opinions of Tristram Shandy, Gentleman*, ed. James A. Work (1759–67; 1940; Indianapolis: Odyssey Press, 1979), 5. Further references to this edition are included in the text.
2. Stephen H. Kellert. *In the Wake of Chaos: Unpredictable Order in Dynamical Systems* (Chicago: University of Chicago Press, 1993), 144.
3. Obviously, a plot cannot and does not infinitely evolve (although one can imagine a Borgesian scenario wherein the task of writing Tristram's life is passed down in an endless succession of authors). Nevertheless, *Tristram Shandy* provides us with the algorithm for an imaginary endless plot.
4. John Milton, *Complete Poems and Major Prose*, ed. Merritt Y. Hughes (Indianapolis: Odyssey Press, 1957), 255.
5. Hans C. Werner also reads *Tristram Shandy* according to a chaos-theory model, but his emphasis is almost exclusively on reader reception. See his *Literary Texts as Nonlinear Patterns: A Chaotics Reading of* Rainforest, Transparent Things, Travesty, *and* Tristram Shandy, Gothenberg Studies in English 75 (Göteborg, Sweden: Acta Universitatis Gothoburgensis, 1999).
6. Kellert, 143.
7. Roland Barthes, *S/Z: An Essay*, trans. Richard Miller (New York: Hill and Wang, 1974) 84; trans. of *S/Z* (Paris: Éditions du Seuil, 1970).
8. R. S. Crane, "The Plot of *Tom Jones*," *The Journal of General Education* 4 (1950): 112–30. Philip Kuberski makes an apt statement about the burgeoning novel genre: "Novelistic characters, like the billiard ball of Newtonian physics, were subject to the laws of action and reaction." See *Chaosmos: Literature, Science, and Theory*, The Margins of Literature (Albany: State University of New York Press, 1994), 18.
9. "Introduction to the Structural Analysis of Narratives," *A Barthes Reader*, ed. Susan Sontag (1966; New York: Hill and Wang, 1982), 266.
10. Both reviews are excerpted in Alan B. Howes, ed., *Sterne: The Critical Heritage* (London: Routledge and Kegan Paul, 1974), 47, 106.

11. Robert Alter, *Partial Magic: The Novel as a Self-Conscious Genre* (Berkeley: University of California Press, 1975), 54.
12. A. A. Mendilow makes an interesting comment about our inability to predict events in *Tristram Shandy*: "Where every episode is presented as in a dramatic present, there can, strictly speaking, be no anticipatory passages or passages of exposition, for there is no fixed line from which to divagate. Such passages when they occur are retrospective or anticipatory only in relation to the time of one incident, and the events the author looks forward to may have been narrated already." See *Time and the Novel* (1952; New York: Humanities Press, 1965), 183.
13. For a detailed argument that Tristram is illegitimate, see John A. Hay, "Rhetoric and Historiography: Tristram Shandy's First Nine Kalendar Months," *Studies in the Eighteenth Century II: Papers Presented at the Second David Nichol Smith Memorial Seminar, Canberra 1970*, ed. R. F. Brissenden (Toronto: University of Toronto Press, 1973), 73–91. Homer Brown also discusses this issue in "Tristram to the Hebrews: Some Notes on the Institution of a Canonic Text," *MLN* 99 (1984): 730.
14. Legally, in fact, as the illegitimate son of Bridget allworthy, Tom would not be Jones, but Allworthy, thus bearing the same name as the squire himself, as I discuss in *The Author's Inheritance: Henry Fielding, Jane Austen, and the Establishment of the Novel* (DeKalb: Northern Illinois University Press, 1998), 87.
15. Wolfgang Iser notes that Tristram's recognition of the impossibility of finding his own beginning depicts "the actual nature of beginnings." See *Laurence Sterne:* Tristram Shandy, Landmarks of World Literature Series (Cambridge: Cambridge University Press, 1988) 5. Significantly, Iser claims that *Tristram Shandy* gives us "a topography of life" (57).
16. Mendilow, 169.
17. Edward N. Lorenz, "Deterministic Nonperiodic Flow," *Journal of the Atmospheric Sciences* 20 (1963): 137; rpt. in *Chaos*, ed. Hao Bai-Lin (1984; Singapore: World Scientific, 1985), 289.
18. Interestingly, Gordon Slethaug locates what I am calling "a strange attractor of death" in in Don DeLillo's *White Noise*: "[D]eath is simultaneously very present and very remote to characters in *White Noise* and certainly cannot be graphed and charted, but, nevertheless, it becomes a strange attractor, generating turbulence and providing pattern." See *Beautiful*

Chaos: Chaos Theory and Metachaotics in Recent American Fiction, The Suny Series in Postmodern Culture (Albany: State University of New York Press, 2000), 148.
19. *Reading for the Plot: Design and Intention in Narrative* (1984; New York: Vintage Books, 1985), 61.
20. James Gleick, *Chaos: Making a New Science* (Viking: New York, 1987), 27.
21. See Iser on Sterne's use of "equivocation" (82–90).
22. Iser point out that, although narrative is metaphorical, Tristram's writing is metonymical (60).
23. See, for example, Robert Alter's chapter "Sterne and the Nostalgia for Reality" in *Partial Magic*, 30–56, and Murray Krieger's essay "The Human Inadequacy of Gulliver, Strephon, and Walter Shandy—and the Barnyard Alternative," *The Classic Vision: The Retreat from Extremity*, vol. 2 of *Visions of Extremity in Modern Literature* (1971; Baltimore: Johns Hopkins University Press, 1973), 255–85.
24. Whether Sterne brought *Tristram Shandy* to completion has been a matter of debate. Wayne C. Booth argues affirmatively in "Did Sterne Complete *Tristram Shandy*," *Modern Philology* 47 (1951): 172–83. See also *The Rhetoric of Fiction*, 2nd edition (Chicago: University of Chicago Press, 1983), 231–32. Louis T. Milic also argues that Sterne "ultimately decided not to write a tenth volume," claiming that he "had exhausted the possibilities of the style of *Tristram Shandy* by the time he finished the ninth volume." See "Information Theory and the Style of *Tristram Shandy*," *The Winged Skull: Papers from the Laurence Sterne Bicentenary Conference*, ed. Arthur H. Cash and John M. Stedmond (Kent, OH: Kent State University Press, 1971), 237. Calling *Tristram Shandy* an example of the genre *il non finito*, Marcia Allentuck, however, argues for an unfinished text: "All of *Tristram Shandy's* parts—from names to noses, and the vast territory in between—stand as wholes, stable, integral, complete, yet never finished, never conclusive." See "In Defense of an Unfinished *Tristram Shandy*," *The Winged Skull*, 153. In the Introduction to his edition of *Tristram Shandy*, Robert Folkenflick discusses the debate, concluding, "The emphases on the finished and unfinished must be taken together for us to see what Sterne was up to." See Introduction, *The Life and Opinions of Tristram Shandy, Gentleman*, by Laurence Sterne (New York: Modern Library, 2004), xxi.

25. See, particularly, Stanley Fish's essay "Normal Circumstances, Literal Language, Direct Speech Acts, the Ordinary, the Everyday, the Obvious, What Goes without Saying, and Other Special Cases," *Is There a Text in This Class: The Authority of Interpretive Communities* (Cambridge: Harvard University Press, 1980), 268–92.
26. Julios Cortázar, *Hopscotch*, trans. Gregory Rabassa (New York: Random House, 1966); trans. of *Rayuela* (Editorial Sudamericana Sociedad Anónima). Milorad Pavić, *Dictionary of the Khazars: A Lexicon Novel in 100,000 Words*, trans. Christina Pribićević-Zorić (New York: Vintage-Random House, 1989). This latter text comes in both a "male" and a "female" edition.
27. *S/Z*, 15.
28. Iser, 76. Mendilow claims that the study of *Tristram Shandy* "could almost serve as a summary of all the problems involved in the consideration of the time factors and values of the novel" (161).
29. Extracts from *The Clockmakers Outcry Against the Author of The Life and Opinions of Tristram Shandy* in Howes, 67–71.
30. Ilya Prigogine and Isabelle Stengers, *Order out of Chaos: Man's New Dialogue with Nature* (Toronto: Bantam Books, 1984), 214.
31. J. T. Fraser, "Time, Infinity, and the World in Enlightenment Thought," *Time, Literature and the Arts: Essays in Honor of Samuel L. Macey*, ed. Thomas R. Cleary, *ELS* Monograph Series (University of Victoria: English Literary Studies, 1994), 200, 201.
32. G. J. Whitrow, *Time in History: The Evolution of Our General Awareness of Time and Temporal Perspective* (Oxford: Oxford University Press, 1988), 3.
33. *Timewatch: The Social Analysis of Time* (Cambridge: Polity Press, 1995), 59.
34. Dorothy Van Ghent, *The English Novel: Form and Function* (1953; New York: Holt, Rinehart and Winston, 1961), 92.
35. Krieger, 279–80.
36. Helene Moglen, *The Philosophical Irony of Laurence Sterne* (Gainesville: The University Presses of Florida, 1975), 57.
37. Moglen, 59.
38. "Tristram Shandy's Consent to Incompleteness: Discourse, Disavowal, Disruption," *Literature and Psychology* 36 (1990): 44–62; rpt. in *Critical Essays on Laurence Sterne*, ed. Melvyn

New (New York: G. K. Hall and Co., 1998), 215, 216. See, also, Iser, who discusses a disjunction between "imagined" and "measurable" time (77–80) and A. A. Mendilow, who discusses a disjunction between "psychological and chronological time" (169–184).

39. *The History of Tom Jones, A Foundling*, introd. Martin C. Battestin, ed. Fredson Bowers (Middletown, CT: Wesleyan University Press, 1975), 76.

40. Jean-Jacques Mayoux argues that "Sterne orders and constructs all times and their relations to ensure the utmost intellectual interest." See "Variations on the Time-sense in *Tristram Shandy*," *The Winged Skull*, 9.

41. Genette, *Narrative Discourse*, 87–88.

42. Genette, *Narrative Discourse*, 87. Of relevance, also, is Bastiaan C. Van Fraassen's observation: "the construction of narrative time is always essentially internal to the text, even when the text gives every sign of wanting to be related to extratextual reality." See "Time in Physical and Narrative Structure," *Chronotypes: The Construction of Time*, ed. John Bender and David E. Wellbery (Stanford: Stanford University Press, 1991), 23.

43. Genette, *Narrative Discourse*, 86. Genette elaborates upon this point in *Narrative Discourse Revisited*, trans. Jane E. Lewin (Ithaca: Cornell University Press, 1988): "But a written narrative, which in that form obviously has no duration, finds its 'reception,' and therefore fully exists, only in an act of performance, whether reading or recitation, oral or silent; and that act has indeed its own duration, but one that varies with the circumstances" (33).

44. In her discussion of postmodern fiction, Ursula Heise comments on the problematic issue of narrative time: " 'time of narration' is a meaningful concept only in precisely those cases in which it forms part of what is narrated—in other words, in those cases in which it is in fact part of the narrated time of the *story*.' " See *Chronoschisms: Time, Narrative, and Postmodernism* (Cambridge: Cambridge University Press, 1997), 151.

45. Krieger, 280.

46. The text indeed seems to exemplify what David Higdon calls a "polytemporal time-shape," in which a writer "freely mixes the various times of the characters, narrator, creator, and reader in such a way that a reader often loses control of all time

references." See *Time and English Fiction* (Totowa, NJ: Rowman and Littlefield, 1977), 11.
47. See, also, Mendilow's incisive analysis of the scene (171–72).
48. John Locke, *An Essay Concerning Human Understanding*, ed. Peter H. Nidditch (1689; Oxford: Clarendon Press, 1982), 186. Moglen provides a nuanced reading of Sterne's treatment of duration (56–61). See, too, Jean-Claude Sallé's argument that Sterne's relativistic concept of duration is a misreading of Locke, a misreading he got from Addison, in "A Source of Sterne's Conception of Time," *Review of English Studies* n.s. 6 (1955): 180–82.
49. In *Narrative Discourse Revisited*, Genette discusses the temporal conversions that occur as we read a text: "To compare the two durations (of story and of reading), one must in reality perform two conversions—from duration of story into length of text, then from length of text into duration of reading" (34).
50. *Narrative Discourse*, 222.
51. Mendilow comments that Sterne "can create an impression of all parts of the story proceeding simultaneously, each at its own pace and in its own direction" (177).
52. Van Fraassen, 24.
53. Paul Harris, online posting, October 13, 1995, International Society for the Study of Time Listserv, October 13, 1995, ISST-L@PSUVM.PSU.EDU. Harris elaborates on his notion of scaling time in "Fractal Faulkner: Scaling Time in *Go Down, Moses*," *Poetics Today* 14 (1993): 625–51.
54. Thomas Weissert, online posting, October 31, 1995, International Society for the Study of Time Listserv, October 31, 1995, ISST-L@PSUVM.PSU.EDU.
55. "From Chaos to Conflict," *Time, Order, Chaos: The Study of Time IX*, ed. J. T. Fraser, Marlene Soulsby, and Alexander Argyros (Madison, CT: International Universities Press, 1998), 12.
56. *Complexification: Explaining a Paradoxical World through the Science of Surprise* (New York: HarperCollins, 1994), 200.
57. Casti, 201–3. When an earlier version of this section appeared in the collection *Disrupted Patterns*, N. Katherine Hayles, in a prefatory essay, offered her own example of chaotic time: "My first encounter with this concept in a scientific context was in a discussion of glasses that crystallize at highly uneven rates. In some parts of the glass crystallization took place in

microseconds; elsewhere (and arbitrarily near to this site) crystallization required many, many years. There was thus no uniform rate at which the glass could be said to crystallize, only fractally complex patterns that covered the spectrum from nanoseconds to centuries." See "Preface: Enlightened Chaos," *Disrupted Patterns: On Chaos and Order in the Enlightenment*, ed. Theodore E. D. Braun and John McCarthy, Internationale Forschungen zur Allgemeinen und Vergleichenden Literaturwissenschaft 43 (Amsterdam-Atlanta, GA: Rodopi, 2000), 3.

58. Michel Serres with Bruno Latour, *Conversations on Science, Culture, and Time*, trans. Roxanne Lapidus (Ann Arbor: University of Michigan Press, 1995); trans. of *Éclairissements* (Éditions François Bourin, 1990), 57.
59. Serres, 58.
60. "A Parodying Novel: Sterne's *Tristram Shandy*," *Laurence Sterne: A Collection of Critical Essays*, ed. John Traugott (Englewood Cliffs, NJ: Prentice-Hall, 1968), 89.

Chapter 3   Narrating the Workings of Memory: Iteration and Attraction in *In Search of Lost Time*

1. A. A. Mendilow, *Time and the Novel*, introd. J. Isaacs (1952; New York: Humanities Press, 1965), 160.
2. "Variations on the Time-sense in *Tristram Shandy*," *The Winged Skull: Papers from the Laurence Sterne Bicentenary Conference*, ed. Arthur H. Cash and John M. Stedmond (Kent, OH: Kent State University Press, 1971), 17.
3. See *In Search of Lost Time*, 6 volumes, trans. C. K. Montcrieff, Terence Kilmartin, Andreas Mayor; rev. trans. D. J. Enright (1992–93; New York: Modern Library, 1998–99) 6: 225. This revised translation draws upon the definitive text of *À la recherché du temps perdu*, which was published by the Bibliothèque de la Pléiade in 1989. Subsequent references to the Modern Language edition are included in the text and preceded by the designation ML. I also include page numbers from the one-volume Quarto Gallimard edition based on the Pléiade edition, *À la recherché du temps perdu*, Jean-Yves Tadié (1987–92; Paris: Éditions Gallimard, 1999), preceded by the designation G: for example (ML 6:225/G 2246). As Richard Bales points out,

"The temptation toward autobiographical interpretation—and sometimes it is strong—needs to be eschewed, and these days it routinely is." See Introduction, *The Cambridge Companion to Proust*, ed. Richard Bales (Cambridge: Cambridge University Press, 2001), 2.

4. Critics may differ on what initiates this quest, however. Jack Jordan argues, "In the first four pages of the novel, Proust has given a metaphorical description of the tabula rasa from which the quest for the lost treasure of the narrator's identity begins." See "The Unconscious" in *Cambridge Companion* 100. William C. Carter regards the event of the parents' yielding to the distraught child as what sets the plot in motion: "[T]he crucial goodnight kiss scene in the *Search* . . . sets in motion the Narrator's lost quest to regain his lost will and become a creative person." See "The Vast Structure of Recollection: from Life to Literature" in *Cambridge Companion*, 28. Strictly speaking, we need a distinction between the protagonist, whose quest to assume a literary vocation the text explores, and the narrator, whom the protagonist becomes once he has assumed his vocation and who "writes" from the vantage point of the quest accomplished. For clarity's sake, however, I use the term "narrator" to signify both of these Proustian creations. For a nuanced discussion of the distinction between narrator, protagonist, and writer, see Brian Rogers, "Proust's Narrator," *Cambridge Companion*, 85–99.

5. Gérard Genette, *Narrative Discourse: An Essay in Method*, trans. Jane E. Lewin (1980; Ithaca: Cornell University Press, 1983). Henri Poincaré's phrase comes from *Science and Hypothesis*, trans. W. J. G. (1905; New York: Dover, 1952), 147; trans. of *La science et l'hypothèse* (Paris: Flammarion, 1902).

6. A quote from Bergson with which Georges Poulet opens *L'Espace proustien* is apropos: " Nous justaposons," dit Bergson, "nos états de conscience de manière à les apercevois simultanément; non plus l'un dans l'autre mais l'un à côté de l'autre; bref, nous projetons le temps dans l'espace." ("We juxtapose," says Bergson, "our states of consciousness in a manner so as to perceive them simultaneously; no longer the one in the other but the one beside the other; in brief, we project time in space." My translation.) See *L'Espace proustien* (Éditions Gallimard, 1963), 9.

7. Several others have applied chaos-theory readings to Proust's text. For example, Emily Zants argues that the structure of *Search* is "constructed according to principles of non-linear dynamical

systems rather than those of traditional Newtonian causal and chronological plot development." See *Chaos Theory, Complexity, Cinema, and the Evolution of the French Novel*, Studies in French Literature 25 (Lewison, NY: Edwin Mellen Press, 1996), 21. However, her analysis focuses on the way in which the reader comes to recognize "self-defining patterns of behavior and feelings" (11) rather than on the way in which the text's structuration manifests disorderly order. See also Patrick Brady's "Does God Play Dice? Deterministic Chaos and Stochastic Chance in Proust's *Recherche*," *Chance, Culture, and the Literary Text*, ed. Thomas M. Kavanaugh, *Michigan Romance Studies* 14 (1994): 133–49. Brady discusses the way in which the text manifests certain qualities common to chaotic systems.

8. Genette defines the iterative thus: "*narrating one time* (or rather: *at one time*) *what happened n times*" (116).
9. *Mathematics and the Unexpected* (Chicago: University of Chicago Press, 1988), 26.
10. Ibid., 25.
11. Poincaré, 186.
12. Critics have often remarked on the chaotic qualities of Proust's text. Joshua Landy, for example, notes that Proust's "style forces an obdurately chaotic material into the merest semblance of order." See "The texture of Proust's novel," *Cambridge Companion*, 128. In discussing the narrator's pathological jealousy, Julia Kristeva explains how "painful chaos give[s] birth to a plot." See *Time and Sense: Proust and the Experience of Literature*, trans. Ross Guberman (New York: Columbia University Press, 1996), 30.
13. James P. Crutchfield, J. Doyne Farmer, Norman H. Packard, and Robert S. Shaw, "Chaos," *Scientific American*, December 1986: 51.
14. Kristeva's description of the text as a "closed spiral" felicitously recalls the structure of the strange attractor (30).
15. Consider the work involved in Kepler's mapping of planetary trajectories, as Ekeland describes it: "The Pulkovo library stores thousands of manuscript pages by Kepler, covered with computations. In the *Astronomia Nova* he rounds off fifteen folio pages of computations by complaining to the reader of having had to do these calculations seventy times before getting the right answer" (5).
16. James Gleick, *Chaos: Making a New Science* (New York: Viking, 1987), 264.

17. Thomas Weissert explains: "Numerical integration must proceed in discrete steps; so not every point along a trajectory can be represented. An infinite number of points along the trajectory [is] lost in the grid of the numerical procedure." See *The Genesis of Simulation in Dynamics: Pursuing the Fermi-Pasta-Ulam Problem* (New York: Springer, 1997), 114.
18. Ibid., *Genesis*, 115.
19. Ibid., *Genesis*, 116. Weissert explains: "The state of the physical system can be measured only to some finite accuracy, the uncertainty of which can lead to randomness in simulation."
20. *The Art of Modeling Dynamic Systems: Forecasting for Chaos, Randomness, and Determinism* (New York: John Wiley and Sons, Inc., 1991), 5.
21. Weissert, *Genesis*, 115.
22. Weissert points out that meaning arises from this very randomness: "There arises an inevitable level of white noise due to this alternating iterative process, one which drowns out the possibility of divining meaning precisely, if at all. But without noise, there can be no meaning." See "Dynamics and Narrative: The Time-Identity Conjugation," *Time, Order, Chaos: The Study of Time IX*, ed. J. T. Fraser, Marlene P. Soulsby, and Alexander J. Argyros (Madison, CT: International Universities Press, 1998), 172.
23. Jean-Yves Tadié comments, "The Combray-Illiers pathway is a memory-screen which conceals Auteuil." See *Marcel Proust: A Life*, trans. Euan Cameron (New York: Penguin, 2000), footnote on p. 3; orig. published by Éditions Gallimard, 1996.
24. Kristeva discusses the possible sources for the madeleine—including pilgrim hats and a character in a George Sand novel in her *Time and Sense*, 5–16.
25. Genette defines the singulative in *Narrative Discourse*, 117–18.
26. See Genette, *Narrative Discourse*, 114.
27. Genette reminds us, "In truth, the iterative itself is always more or less figurative." See *Narrative Discourse Revisted*, trans. Jane Lewin (Ithaca: Cornell University Press, 1988), 23.
28. Genette points out instances of the "pseudo-iterative," whereby scenes are presented in the imperfect but "their richness and precision of detail ensure that no reader can seriously believe that they occur and reoccur in that manner, several times, without any variation" (*Narrative Discourse*, 121). In his defense of the term pseudo-iterative in *Narrative Discourse*

Revisited, Genette drily says, "So let us abandon the attempt to be cautious and let us take the iterative as gospel truth: 'Every Saturday, absolutely the same thing happens' (that is what Proust says)" (23).

29. Of relevance are Jacques Derrida's comments on the iterated utterance in "Signature Event Context": "given the structure of iteration, the intention which animates utterance will never be completely present in itself and its content." See "Signature Event Context," *Margins of Philosophy*, trans. Alan Bass (Chicago: University of Chicago Press, 1982), 326; trans. of *Marges de la Philosophie* (Paris: Les Éditions de Minuit, 1972).

30. At one point, the narrator speaks dismissively of characters in novels whose fates are fixed: "one of those heroes of whom the author, in a tone of indifference which is particularly galling, says to us at the end of a book: 'He very seldom comes up from the country now. He has finally decided to end his days there'" (ML 2: 75). The French wording for the final sentence is as follows: "Il a fini par s'y fixer définitivement, etc." (G 386). The phrase "s'y fixer définitivement" translates literally as "settled there definitely," conveying Proust's concern that novels reinforce the notion of determinate existence.

31. Genette cites this passage in reference to what he calls the "iterism" of the Narrator's "temporal sensitivity" (*Narrative Discourse*, 124).

32. Poincaré, 11.

33. Genette comments that, by "Nom de pays: Le nom" ("Place-Names: The Name,"), "the narrative definitively sets in motion and adopts its pace" (46).

34. For a description of this composition process, see Tadié, 564.

35. Tadié, 193. Tadié's chapter "Le Temps Perdu" deals with the initial conception of the novel (563–99). For a detailed discussion of the evolution of *Search*, see also Marion Schmid, "The birth and development of *A la recherche du temps perdu*," *The Cambridge Companion*, 58–73.

36. Schmid, 67.

37. It is tempting to think that, as with Sterne and *Tristram Shandy*, only death could bring Proust's work on his novel to a close. During his final illness, Proust was actually correcting the manuscripts of *The Captive* (*La Prisonnière*) and *The Fugitive* (*Albertine disparue*), and, on the night before he died, he was still dictating new material. See Tadié, 775–76.

Had more time been granted him, would Proust have continued to expand the middle indefinitely? Schmid concludes that Proust could not: "He would . . . not have been able to make any far-reaching changes to the novel's overall framework; in 1922, this was simply no longer an option because of the state of publication he had reached" (72). That is, the published text, was far enough along that global changes in its overall structure could not be made. Tadié tells us that in spring 1922, Proust told Céleste Albaret, "I have important news. Tonight, I wrote the word 'end' . . . Now, I can die" (762). We must assume that, unlike Sterne with *Tristram Shandy*, Proust had finished *Search*. The text's structuration, however, holds out the tantalizing possibility that the narrative trajectory could continue to visit other sites it its "state space."

38. The initial Modern Library edition (1928) gave this section the apt title "Overture."
39. Tadié notes that, in 1914, "the book was thrown into confusion by the invention of the character of Albertine" (605).
40. Referring to the behavior of the dripping faucet, Robert Shaw notes: "The qualitative type of behavior at a give flow rate, e.g. periodic or chaotic, can depend on initial conditions." (7).
41. Genette, *Narrative Discourse*, 267.

Chapter 4   Narrating the Unbounded: Mrs. Dalloway's Life, Septimus's Death, and Sally's Kiss

1. Virginia Woolf, *Mrs. Dalloway* (1925; San Diego: Harvest-Harcourt Brace and Company, 1981), 79. Further references to the 1981 edition are included in the text.
2. Nancy Topping Bazin discusses Woolf's admiration for Sterne and Proust in *Virginia Woolf and the Androgynous Vision* (New Brunswick: Rutgers University Press, 1973), 44. David Dowling claims that Woolf "learned the tunneling technique" from Proust. See *Mrs. Dalloway: Mapping Streams of Consciousness* (Boston: Twayne Publishers, 1991), 10.
3. "Phases of Fiction," *Collected Essays*, vol. 2 (1929; New York: Harcourt, Brace, and World, 1967), 92.
4. "Phases of Fiction," 93.

5. "Mr. Bennet and Mrs. Brown," *The Captain's Deathbed and Other Essays* (1924; New York: Harcourt, Brace and Company, 1950), 105.
6. "Phases of Fiction," 83.
7. Virginia Woolf, "To Margaret Llewelyn Davies," February 9, 1925, letter 1536 of *A Change of Perspective: The Letters of Virginia Woolf*, ed. Nigel Nicolson and Joanne Trautmann, vol. 3: 1923–28 (London: Hogarth Press, 1977) 166; and Virginia Woolf, "To Vanessa Bell," April 21, 1927, letter 1745 of *Letters*, 365.
8. Although Woolf's so-called stream-of-consciousness technique and the novel's overall setup—one day in one city—would seem to owe something to James Joyce's *Ulysses*, Woolf's opinion of Joyce's novel was mixed at best: "I have read 200 pages so far—not a third; & have been amused, stimulated, charmed[,] interested by the first 2 or 3 chapters—to the end of the Cemetery scene; & then puzzled, bored, irritated, & disillusioned as by a queasy undergraduate scratching his pimples." See *The Diary of Virginia Woolf*, vol. 2: 1920–24, ed. Anne Olivier Bell (London: Hogarth Press, 1978), 188–89. At odds with Woolf's vision of modern fiction was Joyce's overall method, which made readers feel "centred in a self which, in spite of its tremor or susceptibility, never embraces or creates what is outside itself and beyond." See "Modern Fiction," *Collected Essays* 2: 108. Harvena Richter, however, makes an extended argument for the text's indebtedness to *Ulysses* in "The *Ulysses* Connection: Clarissa Dalloway's Bloomsday," *Studies in the Novel* 21 (1989): 305–19.
9. *A Room of One's Own* (1929; San Diego: Harvest-Harcourt Brace Jovanovich, 1957) 79. Woolf argues at one point that "it is fatal for any one who writes to think of their sex" (108).
10. *A Room of One's Own*, 80.
11. *A Room of One's Own*, 85. Although Woolf may be referring obliquely to *Life's Creation*, a 1928 novel written by Marie Stopes under the pseudonym "Mary Carmichael," she also tells her readers at the outset of *Room* that they may call her "Mary Beton, Mary Seton, Mary Carmichael, or by any name you please" (5). She thus discourages a facile identification of Mary Carmichael and *Life's Adventure* with an actual writer and text.

12. *A Room of One's Own*, 96.
13. *Writing beyond the Ending: Narrative Strategies of Twentieth-Century Women Writers*, EVERYWOMAN: Studies in History, Literature, and Culture (Bloomington: Indiana University Press, 1985) 32. According to Rachel Blau DuPlessis: "Breaking the sequence is a critique of narrative, restructuring its orders and priorities precisely by attention to specific issues of female identity and its characteristic oscillation" (x). For a detailed discussion of Woolf's aesthetic manifesto for women, see the chapter "Breaking the Sentence; Breaking the Sequence" (31–46).
14. *A Room of One's Own*, 77.
15. Virginia Woolf, "To Gerald Brenan," May 13, 1923, letter 1388 of *Letters*, vol. 3, 36.
16. *Diary*, 248.
17. *Diary*, 272.
18. Sharon Stockton has discussed Woolf's novel in light of chaos science in "Turbulence in the Text: Narrative Complexity in *Mrs. Dalloway*." *New Orleans Review* 18 (1991): 46–55. However, our arguments go in very different directions. She focuses upon Septimus Smith as a disruptive force that breaks the narrative system.
19. *Diary*, 13.
20. The passage felicitously brings together temporal and spatial trajectories, as the clock's chimes mark off the distance that Dalloway and Whitbread put between themselves and Lady Bruton. Mapping the temporal and/or spatial progression of *Mrs. Dalloway* has preoccupied various critics. Dowling, for example, plots the spatial trajectory of the main characters on an actual street map of London (53–55). Bazin diagrams the novel as a zigzag, with the characters' interactions at each point and the linking elements forming the lines between each (115). Wendy Patrice Williams discusses the novel in terms of a conical structure. See "Falling through the Cone: The Shape of *Mrs. Dalloway* Makes Its Point," *Virginia Woolf: Emerging Perspectives: Selected Papers from the Third Annual Conference on Virginia Woolf*, ed. Mark Hussey and Vara Neverow (New York: Pace University, 1994), 210–15.
21. See "The Terror and the Ecstasy: The Textual Politics of Virginia Woolf's *Mrs. Dalloway*," *Ambiguous Discourse: Feminist Narratology and British Women Writers*, ed. Kathy Mezei (Chapel Hill: University of North Carolina Press, 1996), 171.

22. Denise Delorey points out that "Mrs. Dalloway both is and isn't the center of the book, and the eponymous title at once suggests the concentration of the narrative on one woman and the broader generic quality" of *Mrs. Dalloway*. See "Parsing the Female Sentence: The Paradox of Containment in Virginia Woolf's Narratives," *Ambiguous Discourse*, 100.
23. After expressing her concerns about Mrs. Dalloway's character, Woolf reminded herself that she "can bring innumerable other characters to her support" (*Diary*, 272). It is precisely this mustering of support that I wish to explore. As Nancy Bazin notes, Woolf "tried to capture the essence of a central character through a montage of diverse impressions" (102).
24. "Mr. Bennett and Mrs. Brown," 119. In her 1924 talk, Woolf was refuting Bennett's charge that Georgian novelists "are unable to create characters that are real, true, and convincing" (95).
25. *The Voyage Out* (New York: George H. Doran, 1920), 82, 83.
26. Dowling, 76.
27. For a discussion of the story's publishing history, see Stella McNichol's introduction to *Mrs. Dalloway's Party: A Short Story Sequence*, by Virgnia Woolf, ed. Stella McNichol (New York: Harcourt Brace Jovanovich, 1973), 9–17.
28. Makiko Minow-Pinkey notes that Clarissa thereby serves patriarchal imperatives: "Clarissa accepts the role prescribed by the paternal law, becoming 'the perfect hostess.'" See *Virginia Woolf and the Problem of the Subject* (New Brunswick: Rutgers University Press, 1987), 72.
29. We may wish to consider the implications of the proper name that the title elides. Woolf may be prompting us to think not only of her Clarissa but Samuel Richardson's. Is she hoping to show that her Clarissa, despite her seeming effacement as Mrs. Dalloway, does, like Richardson's Clarissa, have a rich inner life? Is she making a connection between her Clarissa's submission to patriarchal law and Richardson's? For a fascinating discussion of the connections between the two texts, see Iola Groth's published dissertation *The Epistolary Trace: Letters and Transference in Woolf, Austen, and Freud*, diss. University of California, Irvine (Ann Arbor: UMI, 1990), ATT 9005428.
30. In "Mr. Bennet and Mrs. Brown," Woolf makes an analogy between a hostess and a writer: "A convention in writing is not much different from a convention in manners. Both in life and in literature it is necessary to have some means of bridging the

gulf between the hostess and her unknown guest on the one hand, the writer and his unknown reader on the other" (110). Donna K. Reed remarks upon this connection: "Woolf culled from the cooperative psychology underlying the traditional female role an intimate narrative voice that allows readers to participate in her hostess's 'offering to life': to feel while reading the novel an indescribable delight in moments of oneness with others." See "Merging Voices: *Mrs. Dalloway* and *No Place on Earth*," *Comparative Literature* 47 (1995): 129.

31. As many critics have remarked, it is often difficult to know definitively who is speaking. Minow-Pinkey, for example, notes, "Whenever we try to pinpoint the locus of the subject, we get lost in a discursive mist" (58). For Minow-Pinkey, this technique serves as a feminist strategy, as it does for Donna Reed, who argues that "the elusive, intangible character of women's relations to others . . . can be simulated through the narrative style" (127). For Pamela L. Caughie, the technique is a hallmark of a postmodern-like undermining of authority: "Blurring distinctions between characters and between characters and narrator, Woolf makes the source of a thought doubtful, thereby inhibiting our tendency to seek the author's view in the characters or in the narrator." See *Virginia Woolf and Postmodernism: Literature in Quest and Question of Itself* (Urbana: University of Illinois Press, 1991), 74. See also Kathy Mezei, "Free Indirect Discourse, Gender, and Authority in *Emma*, *Howards End*, and *Mrs. Dalloway*," *Ambiguous Discourse*, 81–88.
32. *Diary*, 263.
33. Caughie discusses the deliberately artful nature of the "obtrusive transitions," regarding them as drawing our attention to the "constructedness" of narrative unity (74–75).
34. "Dynamical Discourse Theory," *Time and Society* 4 (1995): 129.
35. *Diary*, 207.
36. Virginia Woolf, "To Gerald Brenan," June 14, 1925, letter 1560 of *Letters*, 189.
37. DuPlessis, 4. Of course, the fact of his death renders Septimus's story conclusive. But, again, I choose to read the text as being "about" Mrs. Dalloway. It is her story—the postcourtship story of a woman—that Woolf wants to avoid concluding.
38. McNichol, 10. McNichol points out in her introduction that " 'The New Dress' was written in 1924 when Woolf was

revising *Mrs. Dalloway* for publication," and "the other five stories [were] written consecutively and probably not later than May 1925" (14).

39. Patricia Matson, 179. Consider, also, Donna Reed's observation: "The narration of the novel fulfills Clarissa's distinctively female quest; from the beginning it 'seeks out' and mingles voices from a broad spectrum of British society, connecting even the lower-class soldier and the society matron who never meet" (128).
40. Minow-Pinkey, 82. See also Sandra Gilbert and Susan Gubar's discussion of "Big Ben history" and its association with "the public world, masculinity, technology, and the war" in *Letters from the Front* (New Haven: Yale University Press, 1989), 23; vol. 3 of *No Man's Land: The Place of the Woman Writer in the Twentieth Century*, 3 vols, 1988–94.
41. The phrase is J. Hillis Miller's. See "*Mrs. Dalloway*: Repetition as the Raising of the Dead," *Critical Essays on Virginia Woolf*, ed. Morris Beja (Boston: G. K. Hall and Company, 1985), 60; rpt. from *Fiction and Repetition: Seven English Novels* (Cambridge: Harvard University Press, 1982). Miller notes: "So fluid are the boundaries between past and present that the reader sometimes has great difficulty knowing whether he is encountering an image from the character's past or something part of the character's immediate experience" (59). For a detailed analysis of the blurring of temporal sequence in the opening of the novel, see Miller, 59–62.
42. Matson notes, "Woolf's writing draws us into the production of meaning, forces us to make connections, and refuses to grant us a position of mastery over the text" (169).
43. *Virginia Woolf and the Languages of Patriarchy* (Bloomington: University of Indiana Press, 1987), 118.
44. If we read *The Voyage Out* backward through *Mrs. Dalloway*, we may see Richard Dalloway's brutal kiss of Rachel as a response to Clarissa's "failure" to satisfy him sexually.
45. "Unmasking Lesbian Passion: The Inverted World of *Mrs. Dalloway*," *Virginia Woolf: Lesbian Readings*, ed. Eileen Barrett and Patricia Cramer, The Cutting Edge: Lesbian Life and Literature (New York: New York University Press, 1997), 154. See also Bazin, who argues that Septimus's "probably guilty feelings concerning his sexual attraction to Evans may well have made him glad to have death end the relationship" (110).

46. See Richard Hughes review "A Day in London Life," *Critical Essays on Virginia Woolf* 14; rpt. from *Saturday Review of Literature* May 16, 1925, 755. Hughes shifts focus from the concerns of the female title character to those of her male suitor.
47. *A Room of One's Own*, 4.
48. In her analysis of *Mrs. Dalloway* and the critics, Laura Smith demonstrates that Clarissa goes from being a "non-person" in critical assessments of the 1920s to a "thwarted lesbian" in contemporary ones. See "Who Do We Think Clarissa Dalloway Is, Anyway? Re-Search Into Seventy Years of Woolf Criticism," *Re: Reading, Re: Writing, Re: Teaching Virginia Woolf: Selected Papers from the Fourth Annual Conference on Virginia Woolf*, ed. Eileen Barret and Patricia Cramer (New York: Pace University Press, 1995), 215–21. Toni McNaron's discussion of her first reading of *Mrs. Dalloway* drives home the way the text was, for many years, read through a grid of compulsory heterosexuality: "Reading *Mrs. Dalloway* for the first time thrilled me beyond measure, though I must confess I completely glossed over the key lesbian scene that now seems central to the entire book. I cannot remember what I told myself from my closet about the significance of Sally Seton's kissing Clarissa, but it was not until I first taught the novel in 1973 that the scene stood out from the overall narrative canvas." See Toni A. H. McNaron, "A Lesbian Reading Virginia Woolf," *Virginia Woolf: Lesbian Readings*, 11.
49. Although the ruling that Radclyffe Hall's *The Well of Loneliness* was pornographic did not occur until 1928 (three years after the publication of *Mrs. Dalloway*), it signifies that the then-existing cultural milieu was not conducive for open discussions of lesbian love. Quentin Bell discusses Woolf's involvement with the Hall trial in his biography of Woolf. See *Virginia Woolf: A Biography*, vol. 2 (New York: Harcourt Brace Jovanovich, 1972), 138–39. Patricia Cramer observes that Woolf's "simultaneous impulses toward self-expression and self-protection is key to decoding . . . the complex, multilayered style for which she is so famous." Cramer, Introduction, *Virginia Woolf: Lesbian Readings*, 118.
50. Miller argues that Clarissa comes to recognize the paradoxical truth that "Only by throwing it away can life be preserved" (68). Dowling points out, however, that, when Clarissa returns to the party, we are suspended between attitudes: "Here the novel teeters between revolution and capitulation. . . . Is

Clarissa endorsing the system by going back to the party? Or is she trying in the best way she can, given her position, to influence those important guests so that they will allow others their space, as she allows Richard his, Peter his, or Sally hers?" (125).
51. The ending of Marleen Gorris's film version (First Look Pictures, 1998), in which Clarissa and Peter, and Sally and Richard pair up for dancing, aims for a closure that the novel does not have. Gorris may be acknowledging that film audiences are less tolerant of deferral than readers.
52. *Diary*, 312. The passage has certain affinities to Proust's discussion of the sketch in *In Search of Lost Time*: "I said to myself that our social existence, like an artist's studio, is filled with abandoned sketches in which we fancied for a moment that we could set down in permanent form our need for a great love, but it did not occur to me that sometimes, if the sketch is not too old, it may happen that we return to it and make of it a wholly different work, and one that is possibly more important than what we had originally planned." See *In Search of Lost Time*, 6 volumes, trans. C. K. Montcrieff, Terence Kilmartin, Andreas Mayor; rev. trans. D. J. Enright (1992–93; New York: Modern Library, 1998–99) 3: 533–34.

Chapter 5 Narrating the Indeterminate: Shreve McCannon in *Absalom, Absalom!*

1. *Voice and Eye in Faulkner's Fiction* (Athens: University of Georgia Press, 1983), 81.
2. *Reading for the Plot: Design and Intention in Narrative* (1984; New York: 1985), 305.
3. Peter Brooks, 308, 304. For a discussion of the text along the lines of Brooks's, see also Karen McPherson, "*Absalom, Absalom!*: Telling Scratches," *Modern Fiction Studies* 33 (1987): 431–50. Joseph R. Reed, Jr., argues that the text "is a narrative about narrative," Shreve and Quentin "replac[ing] the facts they are given with assumptions that better fit their developing design." See *Faulkner's Narative* (New Haven: Yale University Press. 1973), 47.
4. Ian MacKenzie, "Narratology and Thematics," *Modern Fiction Studies* 33 (1987): 543, 544.
5. Susan Sniader Lanser, *Fictions of Authority: Women Writers and Narrative Voice* (Ithaca: Cornell University Press, 1992), 4.

6. Although his overall argument differs from mine, James A. Snead makes a connection between narrative withholding and the suppression of the black blood in "The 'Joint' of Racism: Withholding the Black in *Absalom, Absalom!*," *William Faulkner's* Absalom, Absalom!, ed. Harold Bloom (New York: Chelsea House Publishers, 1987), 129–41. Applying a chaos-science model to *Go Down, Moses*, Paul Harris points out that "the elegant balance struck between order and disorder" in Faulkner's text "is a product and expression of larger historical and ideological forces." See "Fractal Faulkner: Scaling Time in *Go Down, Moses*," *Poetics Today* 14 (1993): 643.
7. James P. Crutchfield, J. Doyne Farmer, Norman H. Packard, and Robert S. Shaw, "Chaos," *Scientific American* December 1986: 47–48.
8. William Faulkner, *Absalom, Absalom!*, The corrected text (1936; 1986; New York: Vintage International, 1990), 211–12. Further references to this edition are included in the text. Italicized words and passages are Faulkner's unless I indicate otherwise.
9. Several critics have argued that Sutpen intends to work against the patrilineal structure of Southern society. For example, Dirk Kuyk, Jr., suggests that Sutpen's design is "not merely to acquire a dynasty but to acquire it so that he could turn it against dynastic society itself"; see *Sutpen's Design: Interpreting Faulkner's* Absalom, Absalom! (Charlottesville, University Press of Virginia, 1990), 17. Deborah Wilson claims that Sutpen has a "resistance to linearity" that "contradict[s] the logic that supports the patriarchal system in which he wished to participate." See " 'A Shape to Fill a Lack': *Absalom, Absalom!* and the Pattern of History," *The Faulkner Journal* 1 (1991): 67. I would argue, however, that it is not Sutpen but the design to which he clings that resists linearity. For a discussion of Sutpen's negation of human unpredictablity, see Robert Dunne, "*Absalom, Absalom!* and the Ripple-Effect of the Past," *The University of Mississippi Studies in English* 10 (1992): 56–66.
10. Cleanth Brooks points out the Sutpen's position would have been secure even if Bon's black blood had been revealed. See *Toward Yoknapatawpha and Beyond* (New Haven: Yale University Press, 1978), 298. Yet, as the passage cited makes clear, Bon represents a threat to Sutpen's own conception of what constitutes Southern gentility.

11. Linda Kauffman argues that the text sets up an opposition between linear time, linked with the male value of the "ledger," and cyclical time, linked with the female value of the "loom." See "Devious Channels of Decorous Ordering: A Lover's Discourse in *Absalom, Absalom!*," *Modern Fiction Studies* 29 (Summer 1983): 183–200.
12. *Faulkner in the University: Class Conferences at the University of Virginia 1957–58*, Frederick L. Gwinn and Joseph L. Blotner, eds. (Charlottesville, VA: University of Virginia Press, 1959), 274.
13. See *S/Z*, trans. Richard Miller (New York: Hill and Wang, 1974) 4; trans. of *S/Z* (Paris: Éditions du Seuil, 1970).
14. Robert Dale Parker argues that Faulkner even undermines the authority of the Chronology and Genealogy; see "The Chronology and Genealogy of *Absalom, Absalom!*: The Authority of Fiction and the Fiction of Authority," *Studies in American Fiction* 14 (1986): 191–98.
15. The definition comes from Hayden White. See "Narrativity in the Representation of Reality," *The Content of the Form* (Baltimore: Johns Hopkins University Press, 1987), 5.
16. See *Faulkner's Revision of Absalom, Absalom!: A Collation of the Manuscript and the Published Book* (Austin: University of Texas Press, 1971). Critics often attempt to provide a rational explanation for Quentin's and Shreve's apparently sudden accession of knowledge by arguing that Henry Sutpen told Quentin of Bon's black blood during the visit that Quentin made to Sutpen's Hundred. See, for example, Michael Millgate, *The Achievement of William Faulkner* (New York: Random House, 1966), 164. See also Cleanth Brooks, *William Faulkner: The Yoknapatawpha Country* (New Haven: Yale University Press, 1963), 424–33, and the entire chapter "The Narrative Structure of *Absalom, Absalom!*" in *Toward Yoknapatawpha and Beyond*, 301–28. Brooks here invents additional dialogue between Henry and Quentin: "If Quentin had merely formed the words 'Charles Bon was your friend—?,' it is easy to imagine Henry's replying: 'More than a friend. My brother.' Faulkner has preferred to leave it to his reader to imagine this or something like it" (322). Hershel Parker argues that it was not what Quentin heard but what he saw at Sutpen's Hundred that made the familial relationship clear, explaining that Quentin realizes then that Bond is Sutpen's descendant: "In

the published texts, Quentin 'remembered how he thought, "The scion, the heir, the apparent (though not obvious)" ' (296)." See "What Quentin Saw 'Out There,' "*Critical Essays on William Faulkner: The Sutpen Family*, ed. Arthur F. Kinney (New York: G. K. Hall, 1996), 276. But even if we assume that the revelation occurred at this time, Faulkner elides the actual scene and leaves it to Shreve and Quentin to invent the interchanges between Sutpen and Henry, as well as Henry and Bon, thus emphasizing the uncertain origin of this crucial piece of knowledge.
17. Thomas Weissert explains that a strange attractor "can be said to exist only in a parametric temporality, which must be mediated by the dynamicist." See "Dynamical Discourse Theory," *Time and Society* 4 (1995): 128–29.
18. Katherine Hayles offers a vivid description of the interaction between the dynamicist and the simulation: "[S]he watches as the screen display generated by the recursion evolves into constantly changing, often unexpected patterns. As the display continues, she adjusts the parameters to achieve different effects. With her own responses in a feedback loop with the computer, she develops an intuitive feeling for how the display and parameters interact. . . . And she is subliminally aware that her interaction with the display could be thought of as one complex system (the behavior described by a set of nonlinear differential equations) interfaced with another (the human neural system) through the medium of the computer." See "Introduction: Complex Dynamics in Literature and Science," *Chaos and Order: Complex Dynamics in Literature and Science*, ed. N. Katherine Hayles (Chicago: University of Chicago Press, 1991), 6.
19. Weissert, "Dynamical Discourse," 125.
20. I should add Native American blood as well. Sutpen's Hundred is part of the Chickasaw land grant. The land for the plantations of Jefferson County came from the cheating and displacement of the Chickasaws.
21. Faulkner admired both Sterne and Proust. In a letter to H. L. Mencken, he mentions that the *American Language Supplement* is "good reading, like Sterne or Swift." See William Faulkner, "To H. L. Mencken," 22 February 1948, *Selected Letters of William Faulkner*, ed. Joseph Blotner (Franklin Center, PA: Franklin Library, 1976), 324. In his biography of Faulkner, Blotner describes some of the points

Faulkner made in an interview with a French doctoral candidate: "When he had read Proust's *À la Recherche du Temps Perdu*, he said, 'This is it!' and wished he had written it himself." See Joseph Blotner, *Faulkner: A Biography* (1 vol.) (1974; New York: Vintage, 1984), 562.

Bibliography

Adam, Barbara. *Timewatch: The Social Analysis of Time.* Cambridge: Polity Press, 1995.
Alcoff, Linda. "Cultural Feminism versus Post-structuralism: The Identity Crisis in Feminist Theory." *The Second Wave: A Reader in Feminist Theory.* Ed. Linda Nicholson. New York: Routledge, 1997. 330–55. Rpt. of "Cultural Feminism vs. Poststructuralism." *Signs* 13 (1988): 405–36.
Allentuck, Marcia. "In Defense of an Unfinished *Tristram Shandy.*" *The Winged Skull: Papers from the Laurence Sterne Bicentenary Conference.* Ed. Arthur H. Cash and John M. Stedmond. Kent, OH: Kent State University Press, 1971. 145–55.
Alter, Robert. *Partial Magic: The Novel as a Self-Conscious Genre.* Berkeley: University of California Press, 1975.
Argyros, Alex. J. *A Blessed Rage for Order: Deconstruction, Evolution, and Chaos.* Ann Arbor: University of Michigan Press, 1991.
Aronofsky, Darren, writer and dir. *Pi.* Artisan Entertainment, 1998.
Aronowitz, Stanley. "The Politics of the Science Wars." *Science Wars.* Ed. Andrew Ross. Durham: Duke University Press, 1996. 202–25.
Bales, Richard, ed. *The Cambridge Companion to Proust.* Cambridge: Cambridge University Press, 2001.
———. Introduction. *The Cambridge Companion to Proust.* Cambridge: Cambridge University Press, 2001.
Barrett, Eileen. "Unmasking Lesbian Passion: The Inverted World of *Mrs. Dalloway. Virginia Woolf: Lesbian Readings.* Ed. Eileen Barrett and Patrician Cramer. The Cutting Edge: Lesbian Life and Literature. New York: New York University Press, 1997. 146–64.
Barrett, Eileen and Patricia Cramer, eds. *Re: Reading, Re: Writing, Re: Teaching Virginia Woolf: Selected Papers from the Fourth Annual Conference on Virginia Woolf.* New York: Pace University Press, 1995.
———. *Virginia Woolf: Lesbian Readings.* The Cutting Edge: Lesbian Life and Literature. New York: New York University Press, 1997.

Barthes, Roland. "Introduction to the Structural Analysis of Narratives." *A Barthes Reader.* Ed. Susan Sontag. 1966; New York: Hill and Wang, 1982. 251–95.

———. *S/Z: An Essay.* Trans. Richard Miller. New York: Hill and Wang, 1974. Trans. of *S/Z* Paris: Éditions du Seuil, 1970.

Bazin, Nancy Topping. *Virginia Woolf and the Androgynous Vision.* New Brunswick: Rutgers University Press, 1973.

Beja, Morris. *Critical Essays on Virginia Woolf.* Boston: G. K. Hall, 1985.

Bell, Quentin. *Virginia Woolf: A Biography.* 2 vols. New York: Harcourt Brace Jovanovich, 1972.

Bender, John and David E. Wellberry. Introduction. *Chronotypes: The Construction of Time.* Ed. John Bender and David E. Wellberry. Stanford: Stanford University Press, 1991.

Berry, Michael. "Chaology: The Emerging Science of Unpredictability." *Royal Institution Proceedings* 61 (1989): 189–204.

Blotner, Joseph. *Faulkner: A Biography* (1 vol.). 1974; New York: Vintage, 1984.

Booth, Wayne C. "Did Sterne Complete *Tristram Shandy.*" *Modern Philology* 47 (1951): 172–83.

———. *The Rhetoric of Fiction.* 2nd Edition. Chicago: University of Chicago Press, 1983.

Brady, Patrick. "Does God Play Dice? Deterministic Chaos and Stochastic Chance in Proust's *Recherche.*" *Chance, Culture, and the Literary Text.* Ed. Thomas M. Kavanaugh. *Michigan Romance Studies* 14 (1994): 133–49.

Braun, Theodore E. D. and John McCarthy, eds. *Disrupted Patterns: On Chaos and Order in the Enlightenment.* Internationale Forschungen zur Allgemeinen und Vergleichenden Literaturwissenschaft 43. Amsterdam-Atlanta, GA: Rodopi, 2000.

———. Foreword. *Disrupted Patterns: On Chaos and Order in the Enlightenment.* Ed. Theodore E. D. Braun and John McCarthy Internationale Forschungen zur Allgemeinen und Vergleichenden Literaturwissenschaft 43. Amsterdam-Atlanta, GA: Rodopi, 2000. v–xiii.

Brooks, Cleanth. *Toward Yoknapatawpha and Beyond.* New Haven: Yale University Press, 1978.

———. *William Faulkner: The Yoknapatawpha Country.* New Haven: Yale University Press, 1963.

Brooks, Peter. *Reading for the Plot: Design and Intention in Narrative.* 1984; New York: Vintage Books, 1985.

Brown, Homer. "Tristram to the Hebrews: Some Notes on the Institution of a Canonic Text." *MLN* 99 (1984): 725–47.
Calvino, Italo. "How I Wrote One of My Books." *Oulipo Laboratory*. Ed. Harry Mathews and Iain White. London: Atlas, 1995. 1–20.
Carter, William C. "The Vast Structure of Recollection: From Life to Literature." *The Cambridge Companion to Proust*. Ed. Richard Bales. Cambridge: Cambridge University Press, 2001. 25–41.
Cash, Arthur H. and John M. Stedmond. *The Winged Skull: Papers from the Laurence Sterne Bicentenary Conference*. Kent, OH: Kent State University Press, 1971.
Casti, John L. *Complexification: Explaining a Paradoxical World through the Science of Surprise*. New York: HarperCollins, 1994.
Caughie, Pamela L. *Virginia Woolf and Postmodernism: Literature in Quest and Question of Itself*. Urbana: University of Illinois Press, 1991.
Cleary, Thomas R. *Time, Literature and the Arts: Essays in Honor of Samuel L. Macey*. ELS Monograph Series. University of Victoria: English Literary Studies, 1994.
Cortázar, Julio. *Hopscotch*. Trans. Gregory Rabassa. New York: Random House, 1966.
Covenay, Peter and Roger Highfield. *The Arrow of Time: A Voyage through Science to Solve Time's Greatest Mystery*. New York: Fawcett Columbine, 1990.
Cramer, Patricia. Introduction to Part 2: "Lesbian Readings of Woolf's Novels." *Virginia Woolf: Lesbian Readings*. Ed. Eileen Barrett and Patrician Cramer. The Cutting Edge: Lesbian Life and Literature. New York: New York University Press, 1997. 117–27.
Crane, R. S. "The Plot of *Tom Jones*." *The Journal of General Education* 4 (1950): 112–30.
Crichton, Michael. *Jurassic Park*. New York: Knopf. 1990.
Crutchfield, James P., J. Doyne Farmer, Norman H. Packard, and Robert S. Shaw. "Chaos." *Scientific American* December 1986: 49–57.
Csicsery-Ronay, Jr., Istvan. "Modeling the Chaosphere: Stanislaw Lem's Alien Communications." *Chaos and Order: Complex Dynamics in Literature and Science*. Ed. N. Katherine Hayles. Chicago: University of Chicago Press, 1991. 244–62.
Delorey, Denise. "Parsing the Female Sentence: The Paradox of Containment in Virginia Woolf's Narratives." *Ambiguous Discourse: Feminist Narratology and British Women Writers*. Ed. Kathy Mezei. Chapel Hill: University of North Carolina Press, 1996. 93–108.

Derrida, Jacques. "Signature Event Context." *Margins of Philosophy.* Trans. Alan Bass. Chicago: University of Chicago Press, 1982. 307–30. Trans. of *Marges de la Philosophie.* Paris: Les Éditions de Minuit, 1972.

Dowling, David. Mrs. Dalloway: *Mapping Streams of Consciousness.* Twayne's Masterworks Studies No. 67. Boston: Twayne Publishers, 1991.

Dunne, Robert. "*Absalom, Absalom!* and the Ripple-Effect of the Past." *The University of Mississippi Studies in English* 10 (1992): 56–66.

DuPlessis, Rachel Blau. *Writing beyond the Ending: Narrative Strategies of Twentieth-Century Women Writers.* EVERYWOMAN: Studies in History, Literature, and Culture. Bloomington: Indiana University Press, 1985.

Ekeland, Ivar. *Mathematics and the Unexpected.* Chicago: University of Chicago Press, 1988. Trans. of *Le Calcul, l'imprévu: Les figures de temps du Kepler à Thom.* Éditions du Seuil, 1984.

Faulkner, William. *Absalom, Absalom!* The corrected text. 1936; 1986; New York: Vintage International, 1990.

———. *Faulkner in the University: Class Conferences at the University of Virginia 1957–58.* Ed. Frederick L. Gwinn and Joseph L. Blotner. Charlottesville, VA: University of Virginia Press, 1959.

———. *Selected Letters of William Faulkner.* Ed. Joseph Blotner. Franklin Center, PA: The Franklin Library, 1976.

Favre, Alexandre, Henri Guitton, Jean Guitton, André Lichnerowicz, and Etienne Wolff. *Chaos and Determinism: Turbulence as a Paradigm for Complex Systems Converging toward Final States.* Trans. Bertram Eugene Schwarzbach. Baltimore: Johns Hopkins University Press, 1995. Trans. of *De la causalité à la finalité. A propos de la turbulence.* Paris: Éditions Maloine, 1988.

Fielding, Henry. *The History of Tom Jones, A Foundling.* Introd. Martin C. Battestin. Ed. Fredson Bowers. Middletown, CT: Wesleyan University Press, 1975.

Fish, Stanley. "Normal Circumstances, Literal Language, Direct Speech Acts, the Ordinary, the Everyday, the Obvious, What Goes without Saying, and Other Special Cases." *Is There a Text in This Class: The Authority of Interpretive Communities.* Cambridge: Harvard University Press, 1980. 268–92.

Folkenflik, Robert. Introduction. *The Life and Opinions of Tristram Shandy, Gentleman.* By Laurence Sterne. New York: Modern Library, 2004. x–xxx.

Fraser, J. T. "From Chaos to Conflict." *Time, Order, Chaos: The Study of Time IX.* Ed. J. T. Fraser, Marlene Soulsby, and

Alexander Argyros. Madison, CT: International Universities Press, 1998. 3–19.

———. "Time, Infinity, and the World in Enlightenment Thought." *Time, Literature and the Arts: Essays in Honor of Samuel L. Macey.* Ed. Thomas R. Cleary. *ELS* Monograph Series. University of Victoria: English Literary Studies, 1994. 192–211.

Fraser, J. T., Marlene Souslby, and Alexander Argyros, eds. *Time, Order, Chaos: The Study of Time IX.* Madison, CT: International Universities Press, 1998. 3–19.

Genette, Gérard. *Narrative Discourse: An Essay in Method.* Trans. Jane E. Lewin. 1980; Ithaca: Cornell University Press, 1983.

———. *Narrative Discourse Revisited.* Trans. Jane Lewin. Ithaca: Cornell University Press, 1988. Trans. of *Nouveau discours du récit.* Paris: Éditions de Seuil, 1983.

Gilbert, Sandra M. and Susan Gubar. *Letters from the Front.* New Haven: Yale University Press, 1989. Vol. 3 of *No Man's Land: The Place of the Woman Writer in the Twentieth Century.* 3 vols. 1988–94.

Glass, Leon and Michael C. Mackey. *From Clocks to Chaos: The Rhythms of Life.* Princeton: Princeton University Press, 1988.

Gleick, James. *Chaos: Making a New Science.* Viking: New York, 1987.

Gorris, Marleen, dir. *Mrs. Dalloway.* First Look Pictures, 1998. Screenplay by Eileen Atkins.

Gross, Paul R. and Norman Levitt. *Higher Superstition: The Academic Left and Its Quarrels with Science.* Baltimore: Johns Hopkins University Press, 1994.

Groth, Iola. *The Epistolary Trace: Letters and Transference in Woolf, Austen, and Freud.* Diss. University of California, Irvine. Ann Arbor: UMI, 1990. ATT 9005428.

Harris, Paul A. "Fractal Faulkner: Scaling Time in *Go Down, Moses.*" *Poetics Today* 14 (1993): 625–51.

———. Online Posting. October 13, 1995. International Society for the Study of Time Listserv. October 13, 1995. <ISST-L@PSUVM.PSU.EDU>.

Hawkins, Harriet. *Strange Attractors: Literature, Culture, and Chaos Theory.* New York. Harvester Wheatsheaf-Prentice Hall, 1995.

Hay, John A. "Rhetoric and Historiography: Tristram Shandy's First Nine Kalendar Months." *Studies in the Eighteenth Century II: Papers Presented at the Second David Nichol Smith Memorial Seminar, Canberra 1970.* Ed. R. F. Brissenden. Toronto: University of Toronto Press, 1973. 73–91.

Hayles, N. Katherine, ed. *Chaos and Order: Complex Dynamics in Literature and Science*. Chicago: University of Chicago Press, 1991.

———. *Chaos Bound: Orderly Disorder in Contemporary Literature and Science*. Ithaca: Cornell University Press, 1990.

———. "Introduction: Complex Dynamics in Literature and Science." *Chaos and Order: Complex Dynamics in Literature and Science*. Ed. N. Katherine Hayles. Chicago: University of Chicago Press, 1991. 1–33.

———. "Preface: Enlightened Chaos." *Disrupted Patterns: On Chaos and Order in the Enlightenment*. Ed. Theodore E. D. Braun and John McCarthy. Internationale Forschungen zur Allgemeinen und Vergleichenden Literaturwissenschaft 43. Amsterdam-Atlanta, GA: Rodopi, 2000. 1–5.

Heise, Ursula. *Chronoschisms: Time, Narrative, and Postmodernism*. Cambridge: Cambridge University Press, 1997.

Higdon, David Leon. *Time and English Fiction*. Totowa, NJ: Rowman and Littlefield, 1977.

Howes, Alan B., ed. *Sterne: The Critical Heritage*. London: Routledge and Kegan Paul, 1974.

Hughes, Richard. "A Day in London Life." *Critical Essays on Virginia Woolf*. Ed. Morris Beja. Boston: G. K. Hall, 1985. 13–15. Rpt. from *Saturday Review of Literature*. May, 16, 1925. 755.

Hunt, Julian C. R. Foreword. *Chaos and Determinism: Turbulence as a Paradigm for Complex Systems Converging toward Final States*. By Alexandre Favre, Henri Guitton, Jean Guitton, André Lichnerowicz, and Etienne Wolff. Trans. Bertram Eugene Schwarzbach. Baltimore: Johns Hopkins University Press, 1995. Trans. of *De la causalité à la finalité. A propos de la turbulence*. Paris: Éditions Maloine, 1988.

Hussey, Mark and Vara Neverow, eds. *Virginia Woolf: Emerging Perspectives: Selected Papers from the Third Annual Conference on Virginia Woolf*. New York: Pace University, 1994.

Iser, Wolfgang. *Laurence Sterne: Tristram Shandy*. Trans. David Henry Wilson. Landmarks of World Literature Series. Cambridge: Cambridge University Press, 1988.

Johnson, Steven. "Strange Attraction." *Lingua Franca: The Review of Academic Life* 6: 3 (1996): 42–50.

Jordan, Jack. "The Unconscious." *The Cambridge Companion to Proust*. Ed. Richard Bales. Cambridge: Cambridge University Press, 2001. 100–116.

Kauffman, Linda. "Devious Channels of Decorous Ordering: A Lover's Discourse in *Absalom, Absalom!*" *Modern Fiction Studies* 29 (Summer 1983): 183–200.

Keller, Evelyn Fox. *Reflections on Gender and Science.* New Haven: Yale University Press, 1985.
Kellert, Stephen H. *In the Wake of Chaos: Unpredictable Order in Dynamical Systems.* Chicago: University of Chicago Press, 1993.
Kinney, Arthur F., ed. *Critical Essays on William Faulkner: The Sutpen Family.* New York: G. K. Hall, 1996.
Koertge, Noretta, ed. *A House Built on Sand: Exposing Postmodernist Myths about Science.* New York: Oxford University Press, 1998.
Krieger, Murray. *The Classic Vision: The Retreat from Extremity.* Vol. 2 of *Visions of Extremity in Modern Literature.* 2 vols. 1971; Baltimore: Johns Hopkins University Press, 1973.
Kristeva, Julia. *Time and Sense: Proust and the Experience of Literature.* Trans. Ross Guberman. New York: Columbia University Press, 1996. Trans. of *Le Temps Sensible: Proust et l'experience littéraire.* Éditions Gallimard, 1996.
Kuberski, Philip. *Chaosmos: Literature, Science, and Theory.* The Margins of Literature. Albany: State University of New York Press, 1994.
Kuyk, Jr., Dirk. *Sutpen's Design: Interpreting Faulkner's* Absalom, Absalom! Charlottesville: University Press of Virginia, 1990.
Landy, Joshua. "The Texture of Proust's Novel." *The Cambridge Companion to Proust.* Ed. Richard Bales. Cambridge: Cambridge University Press, 2001. 117–34.
Lanser, Susan Sniader. *Fictions of Authority: Women Writers and Narrative Voice.* Ithaca: Cornell University Press, 1992.
Livingston, Paisley, ed. *Disorder and Order: Proceedings of the Stanford International Symposium (September 14–16, 1981).* Saratoga, CA: Anma Libri, 1984.
Locke, John. *An Essay concerning Human Understanding.* Ed. Peter H. Nidditch. 1689; Oxford: Clarendon Press, 1982.
Lorenz, Edward N. "Deterministic Nonperiodic Flow." *Journal of the Atmospheric Sciences* 20 (1963): 130–41. Rpt. in *Chaos.* Ed. Hao Bai-Lin. Singapore: World Scientific. 282–93.
Macey, Samuel L. "The Linear and Circular Time Schemes in Sterne's *Tristram Shandy.*" *Notes and Queries* n.s. 36 (1989): 477–79.
MacKenzie, Ian. "Narratology and Thematics." *Modern Fiction Studies* 33 (1987): 535–44.
Marcus, Jane. *Virginia Woolf and the Languages of Patriarchy.* Bloomington: University of Indiana Press, 1987.
Martindale, Colin. "Chaos Theory, Strange Attractors, and the Laws of Literary History." *Empirical Studies of Literature: Proceedings of the*

Second IGEL-Conference, Amsterdam 1989. Ed. Elrud Ibsch, Dick Schram, and Gerard Steen. Amsterdam: Rodopi, 1991. 381–85.

Matson, Patricia. "The Terror and the Ecstasy: The Textual Politics of Virginia Woolf's *Mrs. Dalloway*." *Ambiguous Discourse: Feminist Narratology and British Women Writers*. Ed. Kathy Mezei. Chapel Hill: University of North Carolina Press, 1996. 162–86.

Mayoux, Jean-Jacques. "Variations on the Time-Sense in *Tristram Shandy*." *The Winged Skull: Papers from the Laurence Sterne Bicentenary Conference*. Ed. Arthur H. Cash and John M. Stedmond. Kent, OH: Kent State University Press, 1971. 3–18.

McNaron, Toni A. H. "A Lesbian Reading Virginia Woolf." *Virginia Woolf: Lesbian Readings*. Ed. Eileen Barrett and Patrician Cramer. The Cutting Edge: Lesbian Life and Literature. New York: New York University Press, 1997. 10–20.

McNichol, Stella. Introduction. *Mrs. Dalloway's Party: A Short Story Sequence*. By Virginia Woolf. Ed. Stella McNichol. New York: Harvest-Harcourt Brace Jovanovich, 1973.

McPherson, Karen. "*Absalom, Absalom!*: Telling Scratches," *Modern Fiction Studies* 33 (1987): 431–50.

Mendilow, A. A. *Time and the Novel*. Introd. J. Isaacs. 1952; New York: Humanities Press, 1965.

Mezei, Kathy, ed. *Ambiguous Discourse: Feminist Narratology and British Women Writers*. Chapel Hill: University of North Carolina Press, 1996.

———. "Free Indirect Discourse, Gender, and Authority in *Emma, Howards End*, and *Mrs. Dalloway*." *Ambiguous Discourse: Feminist Narratology and British Women Writers*. Ed. Kathy Mezei. Chapel Hill: University of North Carolina Press, 1996. 66–92.

Milic, Louis T. "Information Theory and the Style of *Tristram Shandy*." *The Winged Skull: Papers from the Laurence Sterne Bicentenary Conference*. Ed. Arthur H. Cash and John M. Stedmond. Kent, OH: Kent State University Press, 1971. 237–46.

Miller, J. Hillis. "*Mrs. Dalloway*: Repetition as the Raising of the Dead." *Critical Essays on Virginia Woolf*. Ed. Morris Beja. Boston: G. K. Hall, 1985. 53–72. Rpt. from *Fiction and Repetition: Seven English Novels*. Cambridge: Harvard University Press, 1982. 176–202, 240.

Millgate, Michael. *The Achievement of William Faulkner*. New York: Random House, 1966.

Milton, John. *Complete Poems and Major Prose*. Ed. Merritt Y. Hughes. Indianapolis: Odyssey Press, 1957.

Minow-Pinkney, Makiko. *Virginia Woolf and the Problem of the Subject*. New Brunswick: Rutgers University Press, 1987.

Moglen, Helene. *The Philosophical Irony of Laurence Sterne*. Gainesville: University Presses of Florida, 1975.
Moretti, Franco. *The Modern Epic: The World System from Goethe to García Márquez*. London: Verso, 1996.
Morin, Edgar. "The Fourth Vision: On the Place of the Observer." Trans. Pierre Saint-Amand. *Disorder and Order: Proceedings of the Stanford International Symposium (Sept. 14–16, 1981)*. Ed. Paisley Livingston. Saratoga, CA: Anma Libri, 1984. 98–108.
Morrison, Foster. *The Art of Modeling Dynamic Systems: Forecasting for Chaos, Randomness, and Determinism*. New York: John Wiley and Sons, Inc., 1991.
Nemesvari, Richard. "Strange Attractors on the Yorkshire Moors: Chaos Theory and *Wuthering Heights*." *The Victorian Newsletter* 92 (1997): 15–21.
New, Melvyn, ed. *Critical Essays on Laurence Sterne*. New York: G. K. Hall, 1998. 127–39.
———. "Sterne and the Narrative of Determinateness." *Eighteenth-Century Fiction* 4 (1992): 315–29. Rpt. in *Critical Essays on Laurence Sterne*. Ed. Melvyn New. New York: G. K. Hall, 1998. 127–39.
Newton, Sir Isaac. *Sir Isaac Newton's Mathematical Principles of Natural Philosophy and His System of the World*. Trans. Andrew Motte. Revised trans. Florian Cajori. 1729; Berkeley: University of California Press, 1946.
Parker, Hershel. "What Quentin Saw 'Out There.'" *Critical Essays on William Faulkner: The Sutpen Family*. Ed. Arthur F. Kinney. New York: G. K. Hall, 1996. 275–78.
Parker, Jo Alyson. *The Author's Inheritance: Henry Fielding, Jane Austen, and the Establishment of the Novel* (DeKalb: Northern Illinois University Press, 1998).
Parker, Robert Dale. "The Chronology and Genealogy of *Absalom, Absalom!*: The Authority of Fiction and the Fiction of Authority." *Studies in American Fiction* 14 (1986): 191–98.
Pavić, Milorad. *Dictionary of the Khazars: A Lexicon Novel in 100,000 Words*. Trans. Christina Pribićević-Zorić. New York: Vintage-Random House, 1989.
Poincaré, Henri. *Science and Hypothesis*. Trans. W. J. G. 1905. New York: Dover, 1952. Trans. of *La science et l'hypothèse*. Paris: Flammarion, 1902.
Pope, Alexander. *Poetry and Prose of Alexander Pope*. Ed. Aubrey Williams. Boston: Houghton Mifflin, 1969.
Poulet, Georges. *L'Espace proustien*. Éditions Gallimard, 1963.

Powers, Richard. *Galatea 2:2*. New York: Farrar, Straus, and Giroux, 1995.

Prigogine, Ilya and Isabelle Stengers. *Order out of Chaos: Man's New Dialogue with Nature*. Toronto: Bantam Books, 1984.

Proust, Marcel. *À la recherché du temps perdu*. Ed. Jean-Yves Tadié. 1987–92; Paris: Éditions Gallimard, 1999.

———. *In Search of Lost Time*. 6 vols. Trans. C. K. Scott Montcrieff, Terence Kilmartin, and Andreas Mayor. Rev. trans. D. J. Enright. 1992–93. New York: Modern Library, 1998–99.

Reed, Donna K. "Merging Voices: *Mrs. Dalloway* and *No Place on Earth*." *Comparative Literature* 47 (1995): 118–35.

Reed, Joseph R., Jr. *Faulkner's Narrative*. New Haven: Yale University Press, 1973.

Rice, Timothy Jackson. *Joyce, Chaos, and Complexity*. Urbana: University of Illinois Press, 1997.

Richter, Harvena. "The *Ulysses* Connection: Clarissa Dalloway's Bloomsday." *Studies in the Novel* 21 (1989): 305–19.

Ricoeur, Paul. *Time and Narrative*. 3 vols. Trans. Kathleen McLaughlin and David Pellauer. Chicago: University of Chicago Press, 1984. Trans. of *Temps et Recit*. Paris: Éditions du Seuil, 1983.

Rogers, Brian. "Proust's Narrator." *The Cambridge Companion to Proust*. Ed. Richard Bales. Cambridge: Cambridge University Press, 2001. 85–99.

Ross, Andrew, ed. *Science Wars*. Durham: Duke University Press, 1996.

Ruppersberg, Hugh M. *Voice and Eye in Faulkner's Fiction*. Athens: University of Georgia Press, 1983.

Sallé, Jean-Claude. "A Source of Sterne's Conception of Time." *Review of English Studies* ns 6 (1955): 180–82.

Schmid, Marion. "The Birth and Development of *À la recherche du temps perdu*." *The Cambridge Companion to Proust*. Ed. Richard Bales. Cambridge: Cambridge University Press, 2001. 58–73.

Serres, Michel with Bruno Latour. *Conversations on Science, Culture, and Time*. Trans. Roxanne Lapidus. Ann Arbor: University of Michigan Press, 1995. Trans. of *Éclairissements*. Éditions François Bourin, 1990.

Shaw, Robert. *The Dripping Faucet as a Model Chaotic System*. Santa Cruz: Aerial Press, 1984.

Shklovsky, Viktor. "A Parodying Novel: Sterne's *Tristram Shandy*." *Laurence Sterne: A Collection of Critical Essays*. Ed. John Traugott. Englewood Cliffs, NJ: Prentice-Hall, 1968. 66–89.

Slethaug, Gordon E. *Beautiful Chaos: Chaos Theory and Metachaotics in Recent American Fiction*. Albany: State University of New York Press, 2000.

Smith, Laura A. "Who Do We Think Clarissa Dalloway Is, Anyway? Re-Search into Seventy Years of Woolf Criticism." *Re: Reading, Re: Writing, Re: Teaching Virginia Woolf: Selected Papers from the Fourth Annual Conference on Virginia Woolf*. Ed. Eileen Barret and Patricia Cramer. New York: Pace University Press, 1995. 215–21.

Snead, James A. "The 'Joint' of Racism: Withholding the Black in *Absalom, Absalom!*." *William Faulkner's* Absalom, Absalom! Ed. Harold Bloom. New York: Chelsea House Publishers, 1987. 129–41.

Sokal, Alan. "Transgressing the Boundaries: An Afterword." *Dissent* 43: 4 (1996): 93–99. Rpt. in *Fashionable Nonsense: Postmodern Intellectuals' Abuse of Science*. Ed. Alan Sokal and Jean Bricmont. New York: Picador USA, 1998. Translation of *Impostures Intellectuelles*. France: Éditions Odile Jacob, 1997. 268–80.

———. "Transgressing the Boundaries: Toward a Transformative Hermeneutics of Quantum Gravity." *Social Text* 46/47 (1996): 217–52. Rpt. in *Fashionable Nonsense: Postmodern Intellectuals' Abuse of Science*. Ed. Alan Sokal and Jean Bricmont. New York: Picador USA, 1998. Translation of *Impostures Intellectuelles*. France: Éditions Odile Jacob, 1997. 212–58.

———. "What the *Social Text* Affair Does and Does Not Prove." *A House Built on Sand: Exposing Postmodernist Myths about Science*. Ed. Noretta Koertge. New York: Oxford University Press, 1998. 9–22.

Sokal, Alan and Jean Bricmont, eds. *Fashionable Nonsense: Postmodern Intellectuals' Abuse of Science*. New York: Picador USA, 1998. Translation of *Impostures Intellectuelles*. France: Éditions Odile Jacob, 1997.

Sterne, Laurence. *Letters of Laurence Sterne*. Ed. Lewis Perry Curtis. Oxford: Clarendon Press, 1967.

———. *The Life and Opinions of Tristram Shandy, Gentleman*. 1759–67; 1940; Ed. James A. Work. Indianapolis: Odyssey Press, 1979.

Stewart, Ian. *Does God Play Dice? The Mathematics of Chaos*. New York: Basil Blackwell, 1989.

Stockton, Sharon. "Turbulence in the Text: Narrative Complexity in *Mrs. Dalloway*." *New Orleans Review* 18 (1991): 46–55.

Stoppard, Tom. *Arcadia*. London: Faber and Faber, 1993.

Tabbi, Joseph and Michael Wutz, eds. *Reading Matters: Narrative in the New Media Ecology*. Ithaca: Cornell University Press, 1997.

Tadié, Jean-Yves. *Marcel Proust: A Life*. Trans. Euan Cameron. New York: Penguin, 2000. Orig. published by Éditions Gallimard, 1996.

Thomas, Calvin. "*Tristram Shandy's* Consent to Incompleteness: Discourse, Disavowal, Disruption." *Literature and Psychology* 36 (1990): 44–62. Rpt. in *Critical Essays on Laurence Sterne*. Ed. Melvyn New. New York: G. K. Hall, 1998. 215–29.

Traugott, John. *Laurence Sterne: A Collection of Critical Essays*. Englewood Cliffs, N.J.: Prentice-Hall, 1968.

Van Fraassen, Bastian C. "Time in Physical and Narrative Structure." *Chronotypes: The Construction of Time*. Ed. John Bender and David E. Wellbery. Stanford: Stanford University Press, 1991. 19–37.

Van Ghent, Dorothy. *The English Novel: Form and Function*. 1953; New York: Holt, Rinehart and Winston, 1961.

Weissert, Thomas P. "Dynamical Discourse Theory." *Time and Society* 4 (1995): 111–33.

———. "Dynamics and Narrative: The Time-Identity Conjugation." *Time, Order, Chaos: The Study of Time IX*. Ed. J. T. Fraser, Marlene P. Soulsby, and Alexander J. Argyros. Madison, CT: International Universities Press, 1998. 163–75.

———. *The Genesis of Simulation in Dynamics: Pursuing the Fermi-Pasta-Ulam Problem*. New York: Springer, 1997.

———. Online Posting. October 31, 1995. International Society for the Study of Time Listserv. October 31, 1995. <ISST-L@PSUVM.PSU.EDU>.

———. "Representation and Bifurcation: Borges's Garden of Chaos Dynamics." *Chaos and Order: Complex Dynamics in Literature and Science*. Ed. N. Katherine Halyes. Chicago: University of Chicago Press, 1991. 222–43.

Werner, Hans C. *Literary Texts as Nonlinear Patterns: A Chaotics Reading of* Rainforest, Transparent Things, Travesty, *and* Tristram Shandy. Gothenberg Studies in English 75. Göteborg, Sweden: Acta Universitatis Gothoburgensis, 1999.

White, Hayden. *The Content of the Form*. Baltimore: Johns Hopkins University Press, 1987.

Whitrow, G. J. *Time in History: The Evolution of Our General Awareness of Time and Temporal Perspective*. Oxford: Oxford University Press, 1988.

Williams, Wendy Patrice. "Falling through the Cone: The Shape of *Mrs. Dalloway* Makes Its Point." *Virginia Woolf: Emerging*

Perspectives: Selected Papers from the Third Annual Conference on Virginia Woolf. Ed. Mark Hussey and Vara Neverow. New York: Pace University, 1994. 210–15.

Wilson, Deborah. "'A Shape to Fill a Lack': *Absalom, Absalom!* and the Pattern of History." *The Faulkner Journal* 1 (1991): 61–81.

Wood, Anthony R. "How Weather Forecasters Got Snowed." *The Philadelphia Inquirer* Janavary. 26, 2000: A1+.

Woolf, Virginia. *A Change of Perspective: The Letters of Virginia Woolf.* Vol. 3: 1923–28. Ed. Nigel Nicolson and Joanne Trautmann. London: Hogarth Press, 1977.

———. *Collected Essays.* Vol. 2. New York: Harcourt, Brace, and World, 1967.

———. *The Diary of Virginia Woolf.* Vol. 2: 1920–24. Ed. Anne Olivier Bell. London: Hogarth Press, 1978.

———. "Modern Fiction." *Collected Essays.* Vol. 2. New York: Harcourt, Brace, and World, 1967. 103–10.

———. "Mr. Bennet and Mrs. Brown." *The Captain's Deathbed and Other Essays.* 1924. New York: Harcourt, Brace and Company, 1950. 94–119.

———. *Mrs. Dalloway.* 1925. San Diego: Harvest-Harcourt Brace and Company, 1981.

———. *Mrs. Dalloway's Party: A Short Story Sequence.* Ed. Stella McNichol. New York: Harvest-Harcourt Brace Jovanovich, 1973.

———. *Orlando: A Biography.* 1928. San Diego: Harvest-Harcourt Brace and Company, 1956.

———. "Phases of Fiction." *Collected Essays.* Vol. 2. 1929. New York: Harcourt, Brace, and World, 1967. 56–102.

———. *A Room of One's Own.* 1929. San Diego: Harvest-Harcourt Brace Jovanovich, 1957.

———. To the Lighthouse. 1927. San Diego: Harvest-Harcourt, 1981.

———. *The Voyage Out.* New York: George H. Doran, 1920.

Work, James A. Introduction. *The Life and Opinions of Tristram Shandy, Gentleman.* By Laurence Sterne. 1940; Indianapolis: Odyssey Press, 1979.

Zants, Emily. *Chaos Theory, Complexity, Cinema, and the Evolution of the French Novel.* Studies in French Literature 25. Lewison, NY: Edwin Mellen Press, 1996.

Index

Adam, Barbara, 47
Alcoff, Linda, 27–28
Allentuck, Marcia, 147
Alter, Robert, 36
Argyros, Alexander, 19, 141
Aronofsky, Darren
   Pi, 19
Aronowitz, Stanley, 143
attractors
   fixed-point, 12, 36
   periodic orbit, 12
   see also strange attractors

baker's transformation, the, 67
Bales, Richard, 151–2
Barrett, Eileen, 107
Barthes, Roland, 27, 34, 119
basin of attraction, 12, 13, 27, 96
Bazin, Nancy Topping, 156, 158, 159, 161
Bergson, Henri, 152
Berry, Michael, 14–15
Booth, Wayne C., 147
bounded randomness, 15, 26, 63, 132, 140
   see also under Proust, Marcel, *In Search of Lost Time*; Woolf, Virginia, *Mrs. Dalloway*
Brady, Patrick, 153
Braun, Theodore E. D., 20
Bricmont, Jean, 20, 143
Brooks, Cleanth, 164, 165
Brooks, Peter, 23, 39, 113
Brown, Homer, 146
Burke, Edmund, 35
butterfly effect, the, 9–10, 18, 44, 138

Carter, William C., 152
Casti, John L., 26, 57
Caughie, Pamela L., 160
chaos, 32–3
   see also deterministic chaos
chaos theory, 1–2, 7–8, 18
   and time, 2, 10, 25
   as pop culture phenomenon, 17–18
   literary applications of, xii–xiii, 2–3, 19–21, 25–9, 131–3: *see also under* chaotic narratives
chaotic narratives, xiii, 131–2
   and connection between form and content, 28–9, 132–3
   definition of, 25–9
   and narrative duration, 25
   and narrative order, 25
   and repeating narrative, 25
   see also under Faulkner, William, *Absalom, Absalom!*; Proust, Marcel, *In Search of Lost Time*; Sterne, Laurence, *The Life and Opinions of Tristram Shandy, Gentleman*; and Woolf, Virginia, *Mrs. Dalloway*
chaotic systems
   atmosphere, the, 8
   "bouncer" toy, 14
   dripping faucet, 9, 12–13, 24, 69, 156
   waterwheel, 40
classical physics, *see* Newtonian physics
Clockmakers Outcry Against the Author of The Life and Opinions of Tristram Shandy, 46

clockwork hegemony, *see under*
   Newtonian physics
Cortázar, Julio
   Hopscotch, 45
Covenay, Peter, 3
Crane, Ronald S., 34
Crichton, Michael
   Jurassic Park, 18, 19
Crutchfield, James P. et al., 10, 13,
   18–19, 114, 135

Delorey, Denise, 159
Derrida, Jacques, 155
determinism, 4–5, 64, 138
deterministic chaos, xi, 1, 7–11
   see also under Proust, Marcel, *In Search of Lost Time*
differential equations, 8, 24, 69, 166
Dowling, David, 93, 156, 158, 162–3
Dunne, Robert, 164
DuPlessis, Rachel Blau, 90, 98
dynamical systems, *see* chaotic narratives, chaotic systems
Dynamical Systems Collective, 135
dynamical systems theory, *see* chaos theory
dynamic of emplotment, 22

Ekeland, Ivar, 5, 17, 64, 135, 153

Farmer, J. Doyne, 135
Faulkner, William, 129–30
   Absalom, Absalom!, 111–30: and
   butterfly effect, 115; and
   determinism, 114–19; and
   reader, 113, 119, 127–8; and
   similarity across scale, 129;
   strange-attractor structure of,
   114, 119, 122–30
   and Proust, Marcel, 166–7
   and Sterne, Laurence, 166
Favre, Alexandre, 7

Fielding, Henry
   The History of Tom Jones, A Foundling, 34, 36, 49
Fish, Stanley, 45
fixed-point attractor, 12, 36
Folkenflik, Robert, 147
fractals, 11, 17, 18, 20, 140
   and similarity across scale, 17, 39
   see also under Faulkner, William, *Absalom, Absalom!*; Proust, Marcel, *In Search of Lost Time*
Fraser, J. T., 10, 23, 46–7, 57

Genette, Gérard, 23, 28, 49–50, 51, 56, 62–3, 71, 83, 143–4, 150, 154–5
Gilbert, Sandra, 161
Gleick, James, 9, 40, 69, 135
Gross, Paul R., 19, 143
Groth, Iola, 159
Gubar, Susan, 161

Harris, Paul A., 57, 144, 164
Hay, John A., 146
Hayles, N. Katherine, 2, 9, 141, 150–1, 166
Heise, Ursula, 149
Hénon, Michele, 17
Higdon, David, 149–50
Highfield, Roger, 3
Hughes, Richard, 162
Hunt, Julian C. R., 3, 4–5

initial conditions, 3, 16
   sensitive dependence on, *see* butterfly effect, the
International Society for the Study of Time, xii
Iser, Wolfgang, 46, 146, 147, 149
iteration, xi, 25, 68–70, 76
iterative mode of frequency, 25
   see also under Proust, Marcel, *In Search of Lost Time*

Johnson, Steven, 21
Jordan, Jack, 152

Kauffman, 165
Keller, Evelyn Fox, 6
Kellert, Stephen, 3, 8, 32, 34
Kenrick, William, 35
Kepler, Johannes, 135, 153
Krieger, Murray, 48, 52
Kristeva, Julia, 153, 154
Kuberski, Philip, 145
Kuyk, Jr., Dirk, 164

Landy, Joshua, 153
Langford, Gerald, 120
Lanser, Susan Sniader, 28, 113
Laplace, Pierre Simon de, 4
  Laplace's demon, 4, 114
Levitt, Norman, 19, 143
linearity, 4, 34
  see also under time
Locke, John, 48, 54
Lorenz, Edward, 4, 8, 10, 39

MacKenzie, Ian, 113
Marcus, Jane, 103
Martindale, Colin, 19
Matson, Patricia, 93, 99, 161
Mayoux, Jean-Jacques, 62, 149
McCarthy, John, 20
McNaron, Toni A. H., 162
McNichol, Stella, 99, 159
McPherson, Karen, 163
Mendilow, A. A., 38, 61–2, 146, 148, 149
Milic, Louis T., 147
Miller, J. Hillis, 161, 162
Millgate, Michael, 165
Milton, John
  Paradise Lost, 33
Minow-Pinkey, Makiko, 101, 159, 160
Moglen, Helene, 48, 150
Moretti, Franco, 29
Morin, Edgar, 11
Morrison, Foster, 70

narratives
  as dynamical systems, 21–5, 132–3
  linear, 35, 36, 38, 62, 132
  structuration, 22
  see also chaotic narratives
Newton, Sir Issac, 5
  Principia Mathematica, 5
Newtonian physics, 3–7, 33, 46–7
  and clockwork universe, 3, 5–7, 8, 32, 47
  and predictability, 3–5, 8
  see also determinism; time
nonlinearity, 2, 9, 18, 131
  in narrative, *see under* Faulkner, William, *Absalom, Absalom!*; Sterne, Laurence, *Tristram Shandy*; Woolf, Virginia, *Mrs. Dalloway*

objectivity, scientific, 3, 6–7, 25
  and observer as producing meaning, 10–11, 126–7

Packard, Norman H., 135
Parker, Hershel, 165–6
Parker, Robert Dale, 165
Pavić, Milorad
  Dictionary of the Khazars, 45
periodic orbit, 12
Poincaré, Henri, 64–5, 76
Pope, Alexander, 66
  An Essay on Man, 7
Prigogine, Ilya, 5, 6, 18, 46
Proust, Marcel, 64, 65, 66, 77, 155–6
  Contre Saine-Beuve, 70
  In Search of Lost Time, 62–3, 64, 65–9, 70–85, 87–8, 163: bounded randomness in, 63, 67, 71, 83; composition process of, 77; and deterministic chaos, 64–5, 66–7; and fictionalized autobiography, 62; and iterative mode of frequency,

Proust, Marcel—*continued*
  63, 68–9, 70–6; and
  similarity across scale, 78–80;
  narrator of, 152; and reader,
  85; strange-attractor
  structure of, 63, 67–8, 71–2,
  75–6, 76–7, 79–83, 85
Jean Santeuil, 70

readers and reading, 2, 3, 21–2, 23,
  24, 25–6, 27, 29, 45–6, 51,
  56, 132
  See also under Faulkner, William,
  Absalom, Absalom!; Sterne,
  Laurence, *Tristram Shandy*;
  Woolf, Virginia, *Mrs.
  Dalloway*
Reed, Donna K., 160
Reed, Jr., Joseph R., 163
Rice, Timothy Jackson, 143
Ricoeur, Paul, 22
Rogers, Brian, 152
Ruppersberg, Hugh, 112

Sallé, Jean-Claude, 150
Schmid, Marion, 77, 155, 156
Second Law of Thermodynamics,
  139
Serres, Michel, 5–6, 58
Shaw, Robert S., 9, 13, 69, 135, 156
Shklovsky, Victor, 58
similarity across scale, *see under*
  fractals
simulation, computer, xi–xii, 1, 11,
  12, 16–17, 22, 23, 24, 69, 127,
  132, 166
Slethaug, Gordon, 146–7
Smith, Laura, 162
Snead, James A., 164
Society for Chaos Theory in
  Psychology and Life Sciences,
  xii, 140
Society for Literature and Science
  (SLS), xii
Sokal, Alan D., 19–20, 143

state space, 12, 16, 23
Stengers, Isabelle, 5, 6, 18, 46
Sterne, Laurence, 39, 45
  *The Life and Opinions of Tristram
  Shandy, Gentleman*, 3, 23,
  26, 31–2, 33–4, 35–46,
  47–59, 61–2, 64, 87–8: and
  determinism, 43–6; and lack
  of an ending, 41–2; and new
  awareness of time, 25, 46,
  48, 56–8; narrative levels in,
  49–51; nonlinearity of, 37–9;
  and reader, 35, 51, 54–5,
  149–50; strange-attractor
  structure of, 32, 33, 36, 37,
  39–41, 52
Stewart, Ian, 7
Stockton, Sharon, 158
Stoppard, Tom
  Arcadia, xi, 19
strange attractors, 11, 12–17, 22–3,
  26, 27, 29, 70, 83, 131
  Lorenz or butterfly strange
  attractor, 15, 24, 34, 39–40
  in narratives, *see under* Faulkner,
  William, *Absalom, Absalom!*;
  Proust, Marcel, *In Search of
  Lost Time*; Sterne, Laurence,
  Tristram Shandy; and Woolf,
  Virginia, *Mrs. Dalloway*
  Rössler or funnel strange
  attractor, 13, 24, 69

Tadié, Jean-Yves, 77, 154, 155, 156
Thomas, Calvin, 48
time
  absolute, 3, 5–6, 10, 25, 32, 46, 47
  chaotic, 2, 10, 25, 46, 56–8,
  150–1
  linear, 5, 57
  relative, 5

Van Franassen, Bastiaan C., 56,
  149
Van Ghent, Dorothy, 47–8

Weissert, Thomas P., 11, 12, 19, 22, 57, 69, 96, 127, 140, 141, 144, 154, 166
Whitrow, G. J., 6, 47
Williams, Wendy Patrice, 158
Wilson, Deborah, 164
Woolf, Virginia
  and Hall, Radclyffe, 162
  Jacob's Room, 89, 92
  and Joyce, James, 157
  "Mr. Bennett and Mrs. Brown," 93, 159–60
  "Mrs. Dalloway in Bond Street," 94
  Mrs. Dalloway, 87–8, 89–110: bounded randomness in, 89, 92; and clock time, 100–1; film version of, 163; and homosexuality, 103–8; nonlinearity of, 101–3; and reader, 103; roving trajectory of focalization in, 91–2, 93, 95–6, 99–100; strange-attractor structure of, 91, 96–100, 102–3, 108–10
  Mrs. Dalloway's Party, 99
  and Proust, Marcel, 88
  Room of One's Own, A, 89, 90, 108
  and Sterne, Laurence, 88
  The Voyage Out, 93, 161
  and woman's writing, 89–91
Work, James, 37

Zants, Emily, 152–3